Finite Element Analysis for Heat Transfer

Theory and Software

Hou-Cheng Huang and Asif S. Usmani

Finite Element Analysis for Heat Transfer

Theory and Software

With 62 Figures

Springer-Verlag
London Berlin Heidelberg New York
Paris Tokyo Hong Kong
Barcelona Budapest

Hou-Cheng Huang, PhD, BSc, MSc
Oxford Instruments UK Ltd., NMR Instruments, Osney Mead,
Oxford OX2 0DX, UK

Asif S. Usmani, PhD, BE, MS
Department of Civil Engineering and Building Science,
The University of Edinburgh, The King's Buildings,
Edinburgh EH9 3JL, UK

Cover illustration: Chapter 7, Figure 5. Sequence of automatically-generated meshes based on a specified target error of 20%.

ISBN-13:978-1-4471-2093-3   e-ISBN-13:978-1-4471-2091-9
DOI: 10.1007/978-1-4471-2091-9

British Library Cataloguing in Publication Data
A catalogue record for this book is available from the British Library

© Springer-Verlag London Limited 1994
Softcover reprint of the hardcover 1st edition 1994

Typesetting: Camera ready produced by authors using LaTeX

69/3830-543210 Printed on acid-free paper

To our families

Preface

This text presents an introduction to the application of the finite element method to the analysis of heat transfer problems. The discussion has been limited to diffusion and convection type of heat transfer in solids and fluids. The main motivation of writing this book stems from two facts. Firstly, we have not come across any other text which provides an introduction to the finite element method (FEM) solely from a heat transfer perspective. Most introductory texts attempt to teach FEM from a structural engineering background, which may distract non-structural engineers from pursuing this important subject with full enthusiasm. We feel that our approach provides a better alternative for non-structural engineers. Secondly, for people who are interested in using FEM for heat transfer, we have attempted to cover a wide range of topics, presenting the essential theory and full implementational details including two FORTRAN programs. In addition to the basic FEM heat transfer concepts and implementation, we have also presented some modern techniques which are being used to enhance the accuracy and speed of the conventional method.

In writing the text we have endeavoured to keep it accessible to persons with qualifications of no more than an engineering graduate. As mentioned earlier this book may be used to learn FEM by beginners, this may include undergraduate students and practicing engineers. However, there is enough advanced material to interest more experienced practitioners.

The first four chapters include a review of the basic heat transfer concepts, the governing equations and a gradual introduction to FEM. The last three chapters present special topics, which are phase change, adaptivity and convection/advection. Finally two appendices give complete details of the programs HEAT2D and HADAPT, with full user instructions and documented examples. HEAT2D is a program capable of steady and transient, linear and nonlinear analyses of diffusive and convective heat transfer. HADAPT is an adaptive version of HEAT2D and includes subroutines for error estimation and unstructured mesh generation of triangular and quadrilateral elements in arbitrary geometries.

The complete source code for both the programs and sample input data files are provided on a floppy disc included with the book.

Swansea, UK   Hou-Cheng Huang, Asif S. Usmani
1993.

Conditions for Program Usage

Contents

Chapter 1

Introduction

1.1 Importance of Numerical Analysis of Heat Transfer

Since the advent of the digital computer, there has been a revolution in the general area of mathematical modelling. Highly sophisticated and detailed analysis of most engineering problems has become possible, where only crude approximations had to be relied upon. For many problems of simple geometry and boundary conditions, exact solutions have existed for decades. Unfortunately most problems of engineering interest involve complicated geometries and boundary conditions. The cost-effectiveness and competitiveness of an engineering product depends greatly on the sophistication of the analysis employed during the design process. This necessitates the use of numerical methods to obtain approximate solutions to real engineering problems.

A large variety of numerical methods have been used to analyse the problems of heat transfer [1]. Some of these methods were used before the power of the digital computer was available, and have thus become redundant. The two numerical methods which have achieved the greatest degree of popularity and success since the arrival of computers have been the finite difference method (FDM) and the finite element method (FEM). FDM has many advantages, not the least of which are its conceptual simplicity and ease of implementation. FDM is applied directly to the differential equation by approximating differentials by appropriate difference expressions. However, the crucial limitation of FDM is that the domain of interest must be divided into rectilinear cells. Therefore, for many problems boundaries have to be approximated by a step like pattern. FEM however, does not suffer from any such limitation, indeed the ability to accurately model the domain boundaries is the major strength of FEM. The domain of interest is divided into a number of simple non-intersecting areas or volumes (*elements*) of variable size (if necessary). The boundaries of these elements may be planar or curved. The field variable in the differential equation is assumed to vary within the element according to a chosen interpolation function. This allows a high degree of flexibility in FEM as accurate results may be obtained by either, using few elements with higher order interpolation, or a larger number of elements with say, linear interpolation. Because of these advantages FEM has been increasingly the numerical method of choice for the analysis of the most complex types of heat transfer problems with unrivalled accuracy.

1.2 Reliability of Finite Element Analysis for Heat Transfer

As mentioned in the previous section FEM represents the most flexible and powerful numerical method for analysing heat transfer problems of general interest. FEM however, is an approximate numerical method and care has to be exercised in setting up a problem for FEM analysis. The quality of the solution obtained depends upon various factors including mainly the distribution of the space discretisation (meshing) throughout the domain, time discretisation for transient problems, proper application of the boundary conditions and selection of suitable material properties. All these aspects of setting up a problem require diligence and experience. Given that proper care has been exercised in setting up the problem for analysis, the results are generally very reliable and provide valuable insight to the designer. However, one needs to be sceptical of all results as errors can creep into the analysis from unforeseeable sources. Therefore it is essential to examine the results carefully and look for anomalies or inconsistencies in them using engineering intuition. The high quality graphical visualisation of results available to the modern user can be of benefit in the interpretation and examination of results. This can also be a source of false security for an unsuspecting user. Therefore it is important to maintain a healthy scepticism and a critical outlook towards the results obtained from all numerical methods.

Different finite element software can be tested using various benchmark solutions available, which are mainly derived from exact analytical solutions of problems with simple geometries. Some benchmark tests for heat transfer problems can be obtained from publications by bodies such as NAFEMS [2].

1.3 Various Heat Transfer Problems

Heat transfer problems constitute a very large class of problems in engineering. These problems span many engineering disciplines including civil, mechanical, chemical, electrical and aeronautical engineering etc. It is beyond the scope of this book to provide a detailed discussion of all the various heat transfer problems. The main aim of this book is to provide basic heat transfer analysis techniques using the finite element method which may be used in a wide variety of applications.

Without reference to any particular application, this text provides means to the reader/user to enable him to solve a large number of different heat transfer problems, some of which may be listed as follows:

1. Heat conduction in solids subject to various boundary conditions:

   - Fixed temperature.
   - Fixed heat flux.
   - Convective heat flux dependent upon the convective heat transfer coefficient.
   - Radiation.

2. Solidification and melting (phase change) problems.

3. Convective heat transfer.

1.4   Objectives and Layout

The main objectives of this book may be listed as follows:

1. Remind the reader of the basic equations that govern the various forms of heat transfer.

2. Explain the background and fundamentals of the finite element method.

3. Describe the general procedure of achieving spatial and temporal discretisation of the the governing equations via the finite element method.

4. Introduce special techniques for dealing with phase change problems.

5. Validate the numerical techniques using standard analytical solutions as benchmark tests.

6. Extend the basic FEM approach to include adaptive analysis based on error estimation.

7. Discuss the difficulties encountered when forced convection problems are solved using standard FEM and how they are overcome.

8. Provide a fully documented set of software with the source code to the reader with test examples covering most topics in the text.

The text in the following chapters has been arranged according to the objectives stated above. The next chapter has been devoted to the establishment of the basic differential equations that govern the different forms of heat transfer and the various boundary conditions that may be applied.

Chapter 3 introduces the reader to the mathematical background of the finite element method following which the basic concepts are explained. The spatial discretisation of the governing equations of heat transfer is then demonstrated using the basic principles established earlier. Discretisation in the time domain for transient (time dependent) problems is discussed in Chapter 4. Various schemes that are normally used in heat transfer analysis are presented.

The special numerical techniques used for solving phase change problems are described in Chapter 5. These techniques are evaluated by comparing with available analytical solutions for freezing and melting.

To obtain an accurate solution, it is sometimes necessary to use very fine meshes in regions of a problem domain where high gradients of the field variable exist. The variation of field variable in the rest of the domain may, on the contrary, be so gentle as to require a very coarse mesh for its adequate resolution. The areas of high gradient may not always be predictable and in the case of transient problems, they may not be restricted to a particular region. Therefore if a fixed fine mesh is used in solving such problems, it may turn out to be unacceptably expensive. For such problems, the technique of adaptive analysis based on estimating the discretisation error provides an economical alternative. This technique is developed for heat transfer problems in Chapter 6.

In the analysis of heat transfer problems where convection is the dominant mode of transport, the conventional Galerkin form of the finite element method proves to be inadequate. Several techniques are used to modify the conventional method to model the convection dominated heat transfer problems successfully. Some of these techniques are discussed in Chapter 7.

The Appendices describe the software included with the text. The main variables used are defined and some key subroutines are listed. The program structures are illustrated schematically. Full user instructions are given for the two programs HEAT2D and HADAPT. Several documented examples are also included for both the programs to familiarise the user with the software. Appendix A presents the program HEAT2D, which may be used for steady and transient, linear and nonlinear analyses of diffusive and convective heat transfer. Phase change problems may also be solved using the enthalpy method or a heat source method. Appendix B introduces the program HADAPT, which is an adaptive version of HEAT2D and includes subroutines for error estimation and unstructured mesh generation of triangular and quadrilateral elements in arbitrary geometries.

References

[1] W.J.Minkowycz, E.M.Sparrow, G.E.Schneider, and R.H.Pletcher, editors. *Handbook of Numerical Heat Transfer.* John Wiley and Sons, Inc., New York, 1988.

[2] J.Barlow and G.A.O.Davies. Selected FE benchmarks in structural and thermal analysis. Technical Report FEBSTA REV 1, NAFEMS, 1986.

Chapter 2

Governing Differential Equations

2.1 Introduction

It is known that as a substance is heated, its temperature increases, thus heat is a kind of energy. This was demonstrated by the British scientist J.P. Joule (1818-1889) in a famous experiment. Heat energy travels from one place to another in several ways. Heat always tends to flow from a region of high temperature to a region of low temperature. There are three basic method by which heat transfer may take place.

If you put one end of a metal bar into the fire, you will eventually notice that the other end of the bar becomes warmer. Heat has travelled through the metal. This form of heat transfer is called *conduction*. The molecules of a hot substance become more energetic and pass on some of their energy to adjacent molecules, which in turn pass on some of their energy to the next molecules, and so on.

In a fluid, such as a gas or a liquid, heat is transferred by a process called *convection*. The fluid in contact with a heat source expands and becomes less dense than the cold fluid surrounding it, and therefore rises. The cold fluid moves in to take its place and in turn expands and rises. Thus, heat is transferred by the movement of the warm fluid in a so-called convection current.

Heat reaches us from the Sun, but there is no substance in space to conduct or convect its heat to us. The heat energy reaches us by *radiation*. In this case, heat is just another form of electromagnetic radiation.

2.2 Conduction

We are not concerned with microscopic interactions within the internal structure. For the purpose of this text, we simply characterize materials by the property of thermal conductivity.

2.2.1   Fourier's Law in Isotropic Materials

We consider at first a plane wall with thickness b, both surfaces of which are kept at different, but constant, temperatures T_{w1} and T_{w2} in steady state (Figure 2.1). The quantity of heat which is caused to flow through the area A of the wall per unit of time due to this temperature difference is called the rate of heat flow, and will be designated by Q. For this quantity of heat, Fourier's law is valid:

$$Q = \frac{k}{b}A(T_{w1} - T_{w2}) \tag{2.1}$$

where k, the *thermal conductivity*, is a property of the substance of which the wall consists. From Equation (2.1) its dimension can easily be derived:

$$k' = \frac{Qb}{A(T_{w1} - T_{w2})} \tag{2.2}$$

thus the dimension of k is

$$Joule/(s)(cm)(°C)$$

or

$$Watt/(cm)(°C)$$

The amount of heat penetrating a unit area of the surface per unit time is called the *heat flux*, q. For this the following equation is valid:

$$q = \frac{Q}{A} = \frac{k}{b}(T_{w1} - T_{w2}) \tag{2.3}$$

As illustrated in Figure 2.1, the temperature within the wall decreases linearly from the value T_{w1} to T_{w2}, if the thermal conductivity is constant. However, this linear relationship does not exist in unsteady heating or cooling process and generally heat flux varies locally and in time. Equation (2.1) may then be applied only to parts of the body of infinitely small dimensions. Therefore, the temperature difference $(T_{w1} - T_{w2})$ in Equation (2.1) is to be replaced by the differential ∂T; the place of the wall thickness b is taken by the differential ∂n, where n is the line normal to the plane of the surface element dA. In this way, for the rate of heat flow dQ through the surface dA, the equation

$$dQ = -kdA\frac{\partial T}{\partial n}$$

is obtained. The minus sign in this equation indicates that the heat flow proceeds in the direction of the temperature drop, that is, in the negative

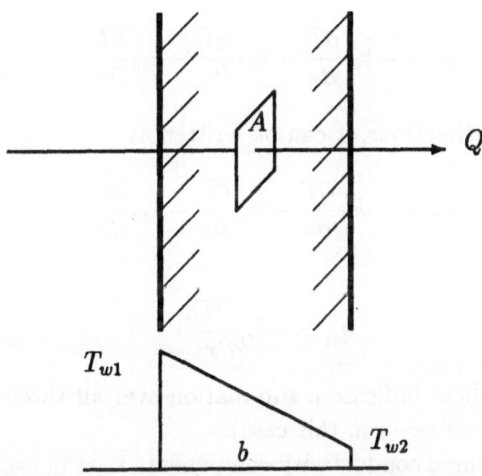

Figure 2.1: Heat Conduction in a Solid Wall

direction of the temperature gradient $\frac{\partial T}{\partial n}$. For the heat flux q_n through the differential element, the equation

$$q_n = -k\frac{\partial T}{\partial n} \tag{2.4}$$

is obtained. Equation (2.4) is the mathematical expression for the basic principle of heat conduction, namely Fourier's Law which states that heat is transferred in proportion to the gradient of temperature.

For a solid described by a Cartesian co-ordinate system, we can then write the heat fluxes in x, y and z directions respectively as follows

$$q_x = -k\frac{\partial T}{\partial x}$$

$$q_y = -k\frac{\partial T}{\partial y}$$

$$q_z = -k\frac{\partial T}{\partial z} \tag{2.5}$$

2.2.2 Fourier's Law in Non-isotropic Materials

The expression (2.5) is valid only for an isotropic material. However, for some materials such as wood, fibrous materials and crystalline substances, the conductivity is directionally dependent. Therefore, the more general non-isotropic description of conduction postulates that any component of the heat flux vector depends upon the temperature gradients in each of three directions. For the x-direction for example

$$q_x = -k_{11}\frac{\partial T}{\partial x} - k_{12}\frac{\partial T}{\partial y} - k_{13}\frac{\partial T}{\partial z} \tag{2.6}$$

for x, y and z (x_i) directions, it can be written as

$$q_{x_i} = -k_{i1}\frac{\partial T}{\partial x} - k_{i2}\frac{\partial T}{\partial y} - k_{i3}\frac{\partial T}{\partial z} \tag{2.7}$$

or simply as

$$q_{x_i} = -k_{ij}\frac{\partial T}{\partial x_j} \tag{2.8}$$

where repeated indices indicate a summation over all three values (summing on repeated suffixes–j in this case).

It is noted that three conductivity components arise in each co-ordinate direction and the thermal conductivity becomes the following tensorial quantity:

$$\mathbf{k} = \begin{bmatrix} k_{11} & k_{12} & k_{13} \\ k_{21} & k_{22} & k_{23} \\ k_{31} & k_{32} & k_{33} \end{bmatrix} \tag{2.9}$$

When the co-ordinate system x, y, z rotates about its origin a new co-ordinate system x', y', z' is obtained. The new conductivity matrix $\mathbf{k}'$ can be written as

$$\mathbf{k}' = \mathbf{P}^T\mathbf{k}\mathbf{P} \tag{2.10}$$

where $\mathbf{P}$ is direction cosine matrix

$$\mathbf{P} = \begin{bmatrix} cos(x\prime, x) & cos(x\prime, y) & cos(x\prime, z) \\ cos(y\prime, x) & cos(y\prime, y) & cos(y\prime, z) \\ cos(z\prime, x) & cos(z\prime, y) & cos(z\prime, z) \end{bmatrix} \tag{2.11}$$

In fact $\mathbf{k}$ is a second order tensor which can be transferred into a diagonal form when an appropriate co-ordinate system is selected. Therefore,

$$\mathbf{P}^T\mathbf{k}\mathbf{P} = \begin{bmatrix} k_\xi & 0 & 0 \\ 0 & k_\eta & 0 \\ 0 & 0 & k_\zeta \end{bmatrix} \tag{2.12}$$

and

$$q_\xi = -k_\xi\frac{\partial T}{\partial \xi}$$

$$q_\eta = -k_\eta\frac{\partial T}{\partial \eta}$$

$$q_\zeta = -k_\zeta \frac{\partial T}{\partial \zeta} \tag{2.13}$$

(ξ, η, ζ) are principal axes of conductivity in the co-ordinate system. k_ξ, k_η and k_ζ are the principal conductivity coefficients. If (ξ, η, ζ) are mutually perpendicular axes then the material is considered as an orthogonal anisotropic material such as wood and fibrous materials. It can also be considered that $k_{ij} = 0(i \neq j)$ and $k_{ii} = k_i$ in the **k** matrix, that is, all non-diagonal terms are zero and the diagonal terms are different, representing different conductivities in different directions. For isotropic case, $k_{ij} = 0(i \neq j)$, and $k_{ii} = k$, that is, all non-diagonal terms in the **k** matrix are zero and the remaining non-zero diagonal ones are equal.

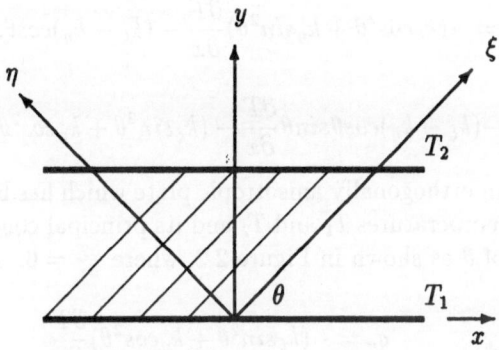

Figure 2.2: Anisotropic material

In order to understand the behaviour of anisotropic materials, we consider the two dimensional case. Assuming that ξ, η are principal axes of conductivity, we have

$$q_\xi = -k_\xi \frac{\partial T}{\partial \xi}$$

$$q_\eta = -k_\eta \frac{\partial T}{\partial \eta}$$

In the x, y co-ordinate system which is obtained by rotating through an angle θ from the $\xi \ \eta$ co-ordinate system, we have

$$\begin{pmatrix} q_x \\ q_y \end{pmatrix} = - \begin{bmatrix} k_{xx} & k_{xy} \\ k_{yx} & k_{yy} \end{bmatrix} \begin{pmatrix} \frac{\partial T}{\partial x} \\ \frac{\partial T}{\partial y} \end{pmatrix} \tag{2.14}$$

and

$$\begin{pmatrix} \xi \\ \eta \end{pmatrix} = \begin{bmatrix} cos\theta & sin\theta \\ -sin\theta & cos\theta \end{bmatrix} \begin{pmatrix} x \\ y \end{pmatrix} \tag{2.15}$$

Therefore,

$$\begin{bmatrix} k_{xx} & k_{xy} \\ k_{yx} & k_{yy} \end{bmatrix} = \begin{bmatrix} cos\theta & -sin\theta \\ sin\theta & cos\theta \end{bmatrix} \begin{bmatrix} k_\xi & 0 \\ 0 & k_\eta \end{bmatrix} \begin{bmatrix} cos\theta & sin\theta \\ -sin\theta & cos\theta \end{bmatrix} \qquad (2.16)$$

It can be obtained that

$$k_{xx} = k_\xi cos^2\theta + k_\eta sin^2\theta$$

$$k_{xy} = k_{yx} = (k_\xi - k_\eta)cos\theta sin\theta$$

$$k_{xx} = k_\xi sin^2\theta + k_\eta cos^2\theta \qquad (2.17)$$

substitution of Equation (2.17) in to Equation (2.14) yields

$$q_x = -(k_\xi cos^2\theta + k_\eta sin^2\theta)\frac{\partial T}{\partial x} - (k_\xi - k_\eta)cos\theta sin\theta\frac{\partial T}{\partial y}$$

$$q_y = -(k_\xi - k_\eta)cos\theta sin\theta\frac{\partial T}{\partial x} - (k_\xi sin^2\theta + k_\eta cos^2\theta)\frac{\partial T}{\partial y} \qquad (2.18)$$

Consider an orthogonally anisotropic plate which has both surfaces fixed at different temperatures T_1 and T_2 and its principal conductivity axes are at an angle of θ as shown in Figure 2.2, where $\frac{\partial T}{\partial x} = 0$. Thus,

$$q_y = -(k_\xi sin^2\theta + k_\eta cos^2\theta)\frac{\partial T}{\partial y} \qquad (2.19)$$

that means

$$k_\theta = k_\xi sin^2\theta + k_\eta cos^2\theta \qquad (2.20)$$

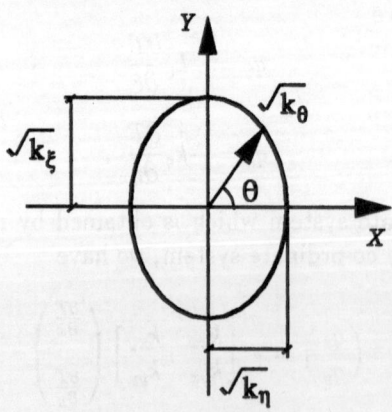

Figure 2.3: Anisotropic heat conductivity coefficients

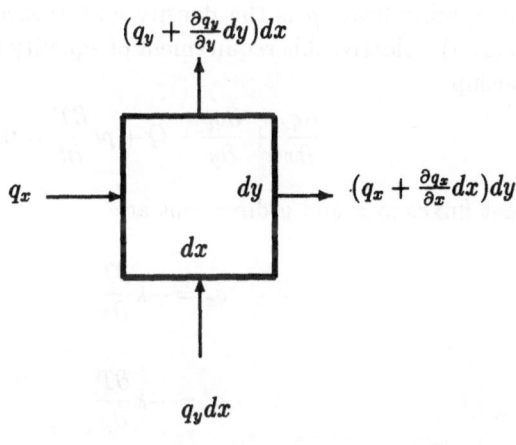

$(q_y + \frac{\partial q_y}{\partial y}dy)dx$

q_x

dy

$(q_x + \frac{\partial q_x}{\partial x}dx)dy$

dx

$q_y dx$

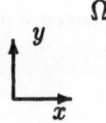

Ω

Figure 2.4: Heat Flow in a 2D element

This is an elliptic function with two principal axes equal to $\sqrt{k_\xi}$ and $\sqrt{k_\eta}$. Figure 2.3 shows that the conductivity coefficient varies for different values of θ. This phenomenon can be shown experimentally by taking a piece of flat glass and a piece of mica of the same shape, and covering one side of both with wax and heating the other side with candles. Few minutes later, it will be found that the shape of the molten wax on the glass is circular while on mica it is ellipsoidal (see [1]).

2.2.3 Governing Equations of Heat Conduction

We now set ourselves the problem of determining the basic equations which govern heat conduction in a solid.

Consider an isotropic material in the two-dimensional system in a domain Ω shown in Figure 2.4. If the heat flowing in the direction of the x and y axes per unit length and in a unit time is denoted by q_x and q_y respectively, the difference between outflow and inflow for an element of size $dxdy$ is given as

$$dy(q_x + \frac{\partial q_x}{\partial x} - q_x) + dx(q_y + \frac{\partial q_y}{\partial y} - q_y)$$

For conservation of heat, this quantity must be equal to the sum of the heat generated in the element in unit time, say, $Qdxdy$ and the heat gained in a unit time due to the temperature change, namely, $-\rho c \frac{\partial T}{\partial t} dx dy$ (where c is the specific heat, ρ is the density and $T(x, y, t)$ is the temperature distribution). Clearly, this requirement of equality leads to the differential relationship

$$\frac{\partial q_x}{\partial x} + \frac{\partial q_y}{\partial y} - Q + \rho c \frac{\partial T}{\partial t} = 0 \qquad (2.21)$$

The heat fluxes in x and y directions are

$$q_x = -k \frac{\partial T}{\partial x}$$

$$q_y = -k \frac{\partial T}{\partial y}$$

Substitution of these heat fluxes into Equation (2.21) produces a higher order differential equation in a single independent variable

$$\frac{\partial}{\partial x}\left(k \frac{\partial T}{\partial x}\right) + \frac{\partial}{\partial y}\left(k \frac{\partial T}{\partial y}\right) + Q - \rho c \frac{\partial T}{\partial t} = 0 \qquad (2.22)$$

In a similar way, the three dimensional equation of heat conduction can be obtained

$$\frac{\partial}{\partial x}\left(k \frac{\partial T}{\partial x}\right) + \frac{\partial}{\partial y}\left(k \frac{\partial T}{\partial y}\right) + \frac{\partial}{\partial z}\left(k \frac{\partial T}{\partial z}\right) + Q - \rho c \frac{\partial T}{\partial t} = 0 \qquad (2.23)$$

it may also be written in vector notation as

$$\nabla \cdot k \nabla T + Q = \rho c \frac{\partial T}{\partial t} \qquad (2.24)$$

This equation is in terms of a Cartesian co-ordinate system (x, y, z). In practice, especially in numerical analysis, another useful system is the cylindrical co-ordinate system. The differential equation of heat conduction in the cylindrical co-ordinate system can be expressed as (see [2])

$$\frac{1}{r}\frac{\partial}{\partial r}\left(rk \frac{\partial T}{\partial r}\right) + \frac{1}{r^2}\frac{\partial}{\partial \theta}\left(k \frac{\partial T}{\partial \theta}\right) + \frac{\partial}{\partial z}\left(k \frac{\partial T}{\partial z}\right) + Q - \rho c \frac{\partial T}{\partial t} = 0 \qquad (2.25)$$

For the general non-isotropic case, the equation of heat conduction in a Cartesian co-ordinate system can be written in tensor notation as

$$\frac{\partial}{\partial x_i}\left(k_{ij} \frac{\partial T}{\partial x_j}\right) + Q = \rho c \frac{\partial T}{\partial t} \qquad (2.26)$$

2.2.4 Initial and Boundary Conditions

In order to solve the partial differential equation of conduction, one needs the specification of *initial condition* at time $t = t_0$ in the domain Ω and of *boundary condition* on the surface Γ for a particular problem.

The initial temperature field must be specified as

$$T(x, y, z, 0) = T_o(x, y, z) \quad \text{in} \quad \Omega \qquad (2.27)$$

There are two typical boundary conditions involved.

In the first condition, the values of temperature at the boundary Γ_T are specified. These values may be constant or be allowed to vary with time, *i.e.*

$$T = T(x, y, z, t) \quad \text{on} \quad \Gamma_T \qquad (2.28)$$

A boundary condition of this form is frequently referred to as a Dirichlet or essential boundary condition.

In the second condition, the values of the heat outflow in the direction n normal to the boundary Γ_q are prescribed as $\bar{q}(x, y, z, t)$. Then we can write

$$- k \frac{\partial T}{\partial n} = \bar{q} \quad \text{on} \quad \Gamma_q \qquad (2.29)$$

This type of boundary condition is often called a Neumann or natural boundary condition.

2.3 Convection

Convective heat transfer takes place when whole groups of molecules move from one place at a certain temperature to another at a different temperature. Therefore, convection is created by fluid flow (Figure 2.5). Imagining a group of particles moving with a given velocity field whose components are (u, v, w) or $(\mathbf{u})$. When calculating the rate at which heat crosses any plane, a convective term of components $(\rho c T u, \rho c T v, \rho c T w)$ or $(\rho c T \mathbf{u})$ must be added to the part due to conduction. Thus the components of the heat flux vector are now

$$q_x = -k \frac{\partial T}{\partial x} + \rho c T u$$

$$q_y = -k \frac{\partial T}{\partial y} + \rho c T v$$

$$q_z = -k \frac{\partial T}{\partial z} + \rho c T w$$

Referring to Equation (2.21) we have

$$\rho c \left(\frac{\partial T}{\partial t} + u \frac{\partial T}{\partial x} + v \frac{\partial T}{\partial y} + w \frac{\partial T}{\partial z} \right) = \nabla \cdot k \nabla T + Q \qquad (2.30)$$

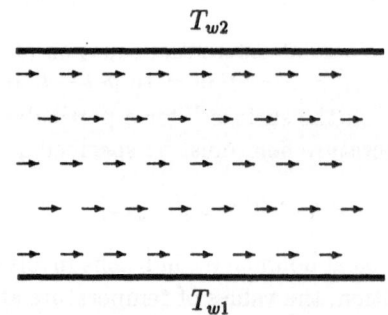

Figure 2.5: Heat Convection in a 2D Flow

This is the equation used for convective heat transfer in a fluid. In most cases of engineering interest which involve fluids convection and diffusion occur simultaneously. Equation (2.30) may be re-written in vector form as

$$\rho c \frac{DT}{Dt} = \nabla \cdot k \nabla T + Q \tag{2.31}$$

where

$$\frac{DT}{Dt} = \frac{\partial T}{\partial t} + u \frac{\partial T}{\partial x} + v \frac{\partial T}{\partial y} + w \frac{\partial T}{\partial z}$$

is called the total or substantial differential. In order to solve Equation (2.31) the same initial and boundary conditions as in Equations (2.27) to (2.29) are required, since the velocity field is assumed to be known.

In the cylindrical co-ordinate system, Equation (2.30) can be written as

$$\rho c \left(\frac{\partial T}{\partial t} + u_r \frac{\partial T}{\partial r} + \frac{u_\theta}{r} \frac{\partial T}{\partial \theta} + u_z \frac{\partial T}{\partial z} \right) =$$

$$\frac{1}{r} \frac{\partial}{\partial r} \left(r k \frac{\partial T}{\partial r} \right) + \frac{1}{r^2} \frac{\partial}{\partial \theta} \left(k \frac{\partial T}{\partial \theta} \right) + \frac{\partial}{\partial z} \left(k \frac{\partial T}{\partial z} \right) + Q \tag{2.32}$$

In general, Equation (2.30) should be coupled with the Navier-Stokes equations. Due to temperature changes in a fluid local density variations are caused. This phenomenon creates buoyancy forces, which induce motion in the fluid. This motion in turn causes heat transfer by convection. Such convection is called *free* or *natural* convection. To model free convection the buoyancy force term must be included in the Navier-Stokes equations (see [3]). In another case, when the motion of the fluid is externally introduced, by a fan for example, then although a small amount of free convection may be present, This phenomenon is known as *forced* convection.

In this text, a given flow field is assumed. It should be noted that the given flow field must satisfy the continuity equation

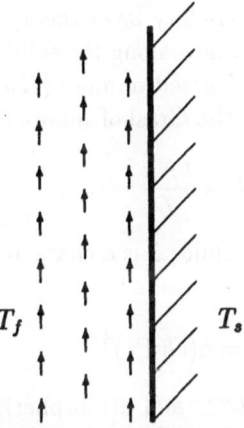

Figure 2.6: Convective Boundary Condition

$$\frac{\partial \rho}{\partial t} + \frac{\partial}{\partial x}(\rho u) + \frac{\partial}{\partial y}(\rho v) + \frac{\partial}{\partial z}(\rho w) = 0 \qquad (2.33)$$

or, in vector notation,

$$\frac{\partial \rho}{\partial t} + \boldsymbol{\nabla}\cdot(\rho\mathbf{u}) = 0$$

We consider the density for the problems relevant to us as constant, *i.e.*, incompressible material, therefore the above equations reduce to a statement of continuity,

$$\frac{\partial u}{\partial x} + \frac{\partial v}{\partial y} + \frac{\partial w}{\partial z} = 0 \qquad (2.34)$$

or,

$$\boldsymbol{\nabla}\cdot\mathbf{u} = 0$$

In many cases, convection normally occurs in heat exchange between fluid and solid boundaries. Therefore, convection may be considered as a kind of boundary condition to the solid domain.

It was Newton who observed that convective heat transfer was a function of the temperature difference between the solid and fluid, T_s and T_f (see Figure 2.6)

$$q = h(T_s - T_f) \qquad (2.35)$$

where q and h are the heat flux and the heat transfer coefficient between the solid and the fluid. Therefore, the boundary condition Equation (2.29) can be expressed as

$$-k\frac{\partial T}{\partial n} = h(T_s - T_f) \qquad \text{on} \quad \Gamma_q \qquad (2.36)$$

To find out the value of h, the boundary layer theory has to be employed, where (2.31) is applied to laminar flow along the solid wall (see [4]). It can be shown that h is related to the Nusselt number (Nu) which is a function of the Prandtl number (Pr) and the Grashof number (Gr). We have

$$h = \frac{kNu}{L} \tag{2.37}$$

where k is the conductivity of the fluid, L is a characteristic dimension and Nu can be simply written as

$$Nu = \alpha(GrPr)^{\frac{1}{4}}$$

For a mean air temperature of $38°C$ and atmospheric pressure with constant wall temperatures, we have (see [5]) $\alpha = 0.59$ for vertical surface and $\alpha = 0.54$ for horizontal surface approximately. The Grashof number (Gr) is defined as

$$Gr = \frac{g\beta(T_s - T_f)L^3}{\nu^2}$$

in which g is the acceleration due to gravity, β is the volumetric coefficient of thermal expansion, and ν is the kinematic viscosity. The Prandtl number (Pr) is defined as

$$Pr = \nu\frac{\rho c_p}{k} = \frac{\mu c_p}{k}$$

where c_p is the specific heat of the fluid and μ is the dynamic viscosity.

2.4   Radiation

In contrast to the mechanisms of conduction and convection, where energy transfer through a material medium is involved, heat may also be transferred by electromagnetic waves. The part of electromagnetic radiation that is propagated as a result of a temperature difference is called *thermal radiation*. It has been established that an ideal thermal radiator, namely, a *blackbody*, will emit energy at a rate proportional to the fourth power of the absolute temperature of the body, *i.e*

$$q = \sigma T^4 \tag{2.38}$$

where σ is the proportionality constant and is called the Stefan-Boltzmann constant with a value of $5.669 \times 10^{-8}W/m^2K^4$. Equation (2.38) governs only radiation emitted from a blackbody. In fact all materials emit a fraction of the blackbody radiation, *i.e.*

$$q = \epsilon\sigma T^4 \tag{2.39}$$

where ϵ is called the emissivity. For a blackbody ϵ is 1. For other surfaces, ϵ is less than 1.

In practice we are normally concerned with radiation between two systems which are at temperatures T_1 and T_2 with $T_1 > T_2$. The surface at T_2 will emit some radiation, but it will receive more radiation from T_1 and hence will increase in temperature. The net heat received by body 2 is

$$F_{12}\epsilon A\sigma(T_1^4 - T_2^4) \tag{2.40}$$

where A is the receiving area of body 2, and F_{12} is a geometry dependent view factor between body 1 and body 2.

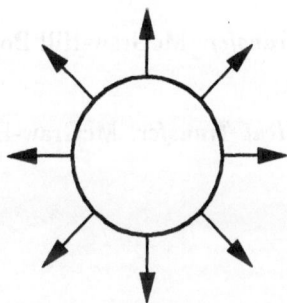

Figure 2.7: Heat Radiation from a Sphere to Space

Since radiation only occurs between two systems, it is apparent that radiation is considered as a boundary condition in the differential equations of heat conduction (see Figure 2.7). A simple radiation problem is encountered when we have a surface at temperature T_1 completely enclosed by a much larger volume maintained at T_2. Then the net radiant exchange in this case can be written as

$$\epsilon_1 A_1\sigma(T_1^4 - T_2^4) \tag{2.41}$$

where ϵ_1 is the emissivity of the enclosed surface. Therefore the boundary condition Equation (2.29) can be written as

$$-k\frac{\partial T}{\partial n} = \epsilon_1\sigma(T_1^4 - T_2^4) \quad \text{on} \quad \Gamma_q \tag{2.42}$$

If we combine Equations (2.29), (2.36) and (2.42) the Neumann boundary condition for the governing equations of the heat transfer Equations (2.24) and (2.30) can be re-written as

$$-k\frac{\partial T}{\partial n} = \bar{q} + h(T_s - T_f) + \epsilon_s\sigma(T_s^4 - T_f^4) \quad \text{on} \quad \Gamma_q \tag{2.43}$$

where ϵ_s is the emissivity of the solid under consideration.

References

[1] C.M.Yu. *Heat Conduction and its Numerical analysis (in Chinese)*. Tsinghua University Press, Beijing, China, 1981.

[2] D.R.Croft and D.G.Lilley. *Heat Transfer Calculations Using Finite Difference Equations*. Applied Science Publishers Ltd, London, U.K., 1977.

[3] L.C.Burmeister. *Convective Heat Transfer*. John Wiley and Sons, New York, U.S.A., 1983.

[4] J.P.Holman. *Heat Transfer*. McGraw-Hill Book Company, New York, U.S.A., 1990.

[5] M.N.Ozisik. *Basic Heat Transfer*. McGraw-Hill Book Company, New York, U.S.A., 1977.

Chapter 3

Finite Element Method

3.1 Introduction

If the heat flow domain and boundary conditions are well posed then the governing equations of heat transfer can be analytically solved but only for the simplest type of problems.

In this section we try to work out the solution of the steady-state heat conduction equation in a rectangular region in which there is no heat generation, the thermal conductivity is constant, and only one of the boundary conditions is nonhomogeneous while the other three boundary conditions are homogeneous. The rectangular region is within $0 \leq x \leq a, 0 \leq y \leq b$, subjected to the boundary conditions shown in Figure 3.1. It may be noted that the surfaces at $x = 0, x = a$, and $y = 0$ are maintained at zero temperature, while the surface at $y = b$ is subjected to a prescribed temperature distribution T_0. The mathematical formulation of this heat conduction problem can be simplified as

$$\frac{\partial^2 T}{\partial x^2} + \frac{\partial^2 T}{\partial y^2} = 0 \qquad (3.1)$$

with

$$T(x, y) = 0 \qquad \text{at} \quad x = 0$$

$$T(x, y) = 0 \qquad \text{at} \quad x = a$$

$$T(x, y) = 0 \qquad \text{at} \quad y = 0$$

$$T(x, y) = T_0 \qquad \text{at} \quad y = b \qquad (3.2)$$

The method of separation of variables is now employed . It is assumed that the temperature $T(x, y)$ can be represented as a product of two functions in the form of

$$T(x, y) = X(x)Y(y) \qquad (3.3)$$

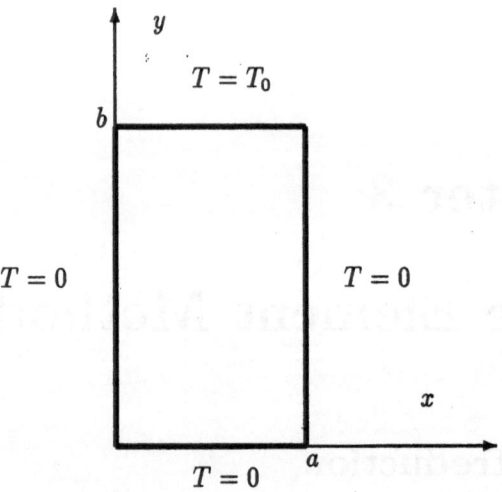

Figure 3.1: Steady-state Heat Conduction in a Rectangular plate

where $X(x)$ is a function of x only and $Y(y)$ a function of y only. Substitution of Equation (3.3) into Equation (3.1) yields

$$\frac{1}{X}\frac{d^2X}{dx^2} = -\frac{1}{Y}\frac{d^2Y}{dy^2} \qquad (3.4)$$

The left-hand side of Equation (3.4) is a function of x only, and the right-hand side is a function of y only. This equality is possible only if both sides are equal to the same constant, say M. Then we have

$$\frac{1}{X}\frac{d^2X}{dx^2} = -\frac{1}{Y}\frac{d^2Y}{dy^2} = M \qquad (3.5)$$

$$\frac{d^2X}{dx^2} = -MX \qquad (3.6)$$

$$\frac{d^2Y}{dy^2} = MY \qquad (3.7)$$

M may be taken as $M = 0$, $M = -\lambda^2 < 0$ or $M = +\lambda^2 > 0$. However, if
$i)M = 0$ then

$$X = Ax + B$$

$$Y = Cy + D$$

$$T(x,y) = (Ax + B)(Cy + D) \qquad (3.8)$$

and

$ii) M = -\lambda^2 < 0$ then

$$X = Acosh\lambda x + Bsinh\lambda x$$

$$Y = Ccos\lambda y + Dsin\lambda y$$

$$T(x,y) = (Acosh\lambda x + Bsinh\lambda x)(Ccos\lambda y + Dsin\lambda y) \qquad (3.9)$$

For both cases their solutions satisfying boundary conditions are the same as (see[1])

$$T \equiv 0$$

which we do not require. Therefore, we choose $M > 0$, that is

iii) $M = +\lambda^2 > 0$ then

$$X = Acos\lambda x + Bsin\lambda x \qquad (3.10)$$

with boundary conditions of

$$X = 0 \qquad \text{at} \quad x = 0$$

$$X = 0 \qquad \text{at} \quad x = a \qquad (3.11)$$

where the boundary conditions (3.11) are obtained from the boundary conditions (3.2) by substituting $T(x,y) = X(x)Y(y)$ and noting that the $Y(y)$ should not vanish at $x = 0$ and $x = a$. Therefore,

$$X(x) = sin\lambda_n \qquad (3.12)$$

where

$$\lambda_n = n\pi/a \qquad (3.13)$$

and constant B is omitted. The solution of Equation (3.7) for the Y separation is taken as

$$Y = Csinh\lambda_n y + Dcosh\lambda_n y \qquad (3.14)$$

with the boundary conditions of

$$Y = 0 \qquad \text{at} \quad y = 0$$

where the boundary condition is obtained from the boundary condition (3.2) by substituting $T(x,y) = X(x)Y(y)$ and noting that $X(x)$ should not vanish at $Y = 0$. Therefore, we have $D = 0$, then the solution for $Y(y)$ is taken as

$$Y(y) = sinh\lambda_n y \qquad (3.15)$$

where constant C is omitted.

It should be noted that the solution for $T(x,y)$ is a product of these two separation functions in the form $T(x,y) = X(x)Y(y) = sin\lambda_n xsinh\lambda_n y$, in this case we do not have a single solution for the separation functions but many solutions for each consecutive value $\lambda_n, n = 1, 2, 3, ...$ Therefore, the complete solution for the temperature should be taken as a linear sum of all these permissible solutions in the form

$$T(x, y) = \sum_{n=1}^{\infty} C_n sinh\lambda_n y sin\lambda_n x \qquad (3.16)$$

where the C_n's are the unknown expansion coefficients which are yet to be determined. We note that the solution given by Equation (3.16) satisfies the differential equation of heat conduction (3.1) and the three homogeneous boundary condition in Equation (3.2). But it does not yet satisfy the non-homogeneous boundary condition. Therefore, the unknown coefficients C_n's can be determined by constraining the above solution to satisfy the non-homogeneous boundary condition in Equation (3.2). If Equation (3.16) is the solution of the above heat conduction problem it should also satisfy the non-homogeneous boundary condition, that is,

$$T_0 = \sum_{n=1}^{\infty} C_n sinh\lambda_n b sin\lambda_n x \qquad in \quad 0 \leq x \leq a \qquad (3.17)$$

We now multiply both sides of Equation (3.17) by $sin\lambda_m x$ and integrate it from $x = 0$ to $x = a$,

$$\int_0^a T_0 sin\lambda_m x' dx' = \sum_{i=1}^{\infty} C_n sinh\lambda_n b \int_0^a sin\lambda_m x' sin\lambda_n x' dx' \qquad (3.18)$$

The orthogonality property of sine waves is utilized in these integrations. That is

$$\int_0^a sin\lambda_m x' sin\lambda_n x' dx' = 0 \qquad for \quad \lambda_n \neq \lambda_m \qquad (3.19)$$

$$\int_0^a sin\lambda_m x' sin\lambda_n x' dx' = \frac{a}{2} \qquad for \quad \lambda_n = \lambda_m \qquad (3.20)$$

all the terms in the summation on the right-hand side of Equation (3.18) vanish except the term $\lambda_n = \lambda_m$. Therefore we have

$$\int_0^a T_0 sin\lambda_n x' dx' = \frac{a}{2} C_n sinh\lambda_n b \qquad (3.21)$$

integrating the left-hand side yields

$$\frac{aT_0}{n\pi}[-(-1)^n + 1] = \frac{a}{2} C_n sinh\lambda_n b \qquad (3.22)$$

and

$$C_n = \frac{2T_0[-(-1)^n + 1]}{n\pi sinh\lambda_n b} \qquad (3.23)$$

The substitution of Equation (3.23) into Equation (3.16) finally gives the desired solution of the above heat conduction problem as

$$T(x,y) = \frac{2T_0}{\pi} \sum_{n=1}^{\infty} \frac{[-(-1)^n + 1]}{n} \frac{sinh\lambda_n y}{sinh\lambda_n b} sin\lambda_n x \qquad (3.24)$$

or

$$T(x,y) = \frac{4T_0}{\pi} \sum_{k=1}^{\infty} \frac{1}{2k+1} \frac{sinh\frac{(2k+1)\pi y}{a}}{sinh\frac{(2k+1)\pi b}{a}} sin\frac{(2k+1)\pi x}{a} \qquad (3.25)$$

It can be seen that the analytical solution of the simplest problem is not simple. Furthermore, a numerical procedure is still required to evaluate the above expression. To overcome such difficulties and to enlist the aid of the most powerful tool developed in this century - the digital computer - it is desirable to recast the problem in a purely algebraic form, involving only the basic arithmetic operations. To achieve this, various forms of *discretisation* of the continuum problem defined by the differential equations can be used. In the continuum problem the solution is satisfied at all points in the problem region. The discretised form of the problem only requires the solution to be satisfied at a finite number of points in the region. In the remainder of the region appropriate interpolations may be used. Of the various forms of discretisation which are possible, one of the simplest is the *finite difference method* which has been utilized for many decades. Today, the *finite element method* developed in the 1960's is the most attractive tool for the approximate solution of differential equations. The finite element formulations can be derived from several different approaches such as variational principles and weighted residuals.

3.2 Variational Principle and Rayleigh-Ritz Method

Some physical problems can be stated directly in the form of a variational principle which consists of determining the function which makes a certain integral statement, called *functional*, stationary. However, the form of the variational principle is not always obvious and, indeed, such a principle does not exist for many continuum problems for which well-defined differential equations may be formulated. Fortunately, a variational principle is available for the differential equation of heat conduction. Consider now the situation of two space dimensions and the functional defined by [2]

$$\Pi(T) = \int_{\Omega} \left[\frac{k}{2}(\frac{\partial T}{\partial x})^2 + \frac{k}{2}(\frac{\partial T}{\partial y})^2 - QT \right] d\Omega + \int_{\Gamma_q} \bar{q} T \, d\Gamma \qquad (3.26)$$

where k, the conductivity, and Q, the heat generation, are functions of position only. We further define the boundary curve as $\Gamma = \Gamma_q + \Gamma_T$. The admissible functions in the variational process will consist of those function satisfying the condition $T = \bar{T}$ on Γ_T.

It can be shown that Equation (3.26) is equivalent to the following differential equation

$$\frac{\partial}{\partial x}(k\frac{\partial T}{\partial x}) + \frac{\partial}{\partial y}(k\frac{\partial T}{\partial y}) + Q = 0 \qquad (3.27)$$

with the boundary conditions of

$$T = T_0 \qquad \text{on} \quad \Gamma_T \qquad (3.28)$$

which has been satisfied in advance and

$$-k\frac{\partial T}{\partial n} = \bar{q} \qquad \text{on} \quad \Gamma_q \qquad (3.29)$$

Using the variational operation, we have

$$\delta\Pi(T) = \int_\Omega \left[\frac{k}{2}(\frac{\partial T}{\partial x})\frac{\partial}{\partial x}(\delta T) + \frac{k}{2}(\frac{\partial T}{\partial y})\frac{\partial}{\partial y}(\delta T) - Q\delta T \right] d\Omega + \int_{\Gamma_q} \bar{q}\delta T d\Gamma = 0 \qquad (3.30)$$

We may rewrite Equation (3.30), using Green's lemma as follows

$$\int_\Omega A\frac{\partial B}{\partial x}dxdy = -\int_\Omega \frac{\partial A}{\partial x}Bdxdy + \oint_\Gamma ABn_x d\Gamma$$

$$\int_\Omega A\frac{\partial B}{\partial y}dxdy = -\int_\Omega \frac{\partial A}{\partial y}Bdxdy + \oint_\Gamma ABn_y d\Gamma \qquad (3.31)$$

where A and B are suitable differential functions, n_x and n_y are the direction cosines of the outward normal n to the closed curve Γ surrounding an area Ω in the (x,y) plane. The integration around Γ is made in an anticlockwise direction and it should noted that

$$n_x\frac{\partial A}{\partial x} + n_y\frac{\partial A}{\partial y} = \frac{\partial A}{\partial n} \qquad (3.32)$$

Applying Equations (3.31) and (3.32) to Equation (3.30), we obtain
$$\delta\Pi(T) =$$

$$\int_{\Gamma_T+\Gamma_q} k\frac{\partial T}{\partial n}\delta T d\Gamma - \int_\Omega \left[\frac{\partial}{\partial x}(k\frac{\partial T}{\partial x}) + \frac{\partial}{\partial y}(k\frac{\partial T}{\partial y}) + Q \right] \delta T d\Omega + \int_{\Gamma_q} \bar{q}\delta T d\Gamma = 0 \qquad (3.33)$$

and since $\delta T = 0$ on Γ_T, therefore,

$$\delta\Pi(T) = \int_{\Gamma_q} (k\frac{\partial T}{\partial n} + \bar{q})\delta T d\Gamma - \int_\Omega \left[\frac{\partial}{\partial x}(k\frac{\partial T}{\partial x}) + \frac{\partial}{\partial y}(k\frac{\partial T}{\partial y}) + Q \right] \delta T d\Omega = 0 \qquad (3.34)$$

where δT is arbitrary, thus Equations (3.27) and (3.29) are necessary when Π is stationary. We call the differential equation (3.27), the *Euler-Lagrange equation* and the boundary condition (3.29), *natural*.

It can be seen that instead of solving the differential equation(Euler-Lagrange equation) directly, we can minimize its functional to approach the problem of two-dimensional steady state heat conduction with the essential and natural boundary conditions on Γ_T and Γ_q

Let us now consider a specific problem, which is a long cylinder of inner radius $r = a$ and outer radius $r = b$. A fixed flux of $q = -U$ per unit length is applied on the inner surface of the cylinder whereas at the outer edge the temperature is fixed at $T_0 = 0$. The governing partial differential equation for this problem is

$$\frac{1}{r}\frac{\partial}{\partial r}(rk\frac{\partial T}{\partial r}) = 0 \tag{3.35}$$

with the boundary conditions of

$$T(r) = 0 \qquad \text{at} \quad r = b$$

$$k\frac{\partial T}{\partial n} - U = 0 \qquad \text{at} \quad r = a \tag{3.36}$$

The classical analytical solution to the heat conduction problem is presented by Carslaw and Jaeger [3] which has the form of

$$T(r) = T_0 + q\frac{a}{k}ln(\frac{r}{b}) = -\frac{Ua}{k}ln(\frac{r}{b}) \tag{3.37}$$

The functional to this problem can be written as follows

$$\Pi(T) = \frac{1}{2}\int_0^{2\pi}\int_a^b k(\frac{\partial T}{\partial r})^2 rdrd\theta - \int_0^{2\pi} UTa d\theta \tag{3.38}$$

We now consider the Rayleigh-Ritz method which uses *trial functions* $N(r)_n$ to approximate the real solution $T(r)$ in the functional. That is

$$\bar{T}(r) = C_0 + \sum_{n=1}^{\infty} C_n N(r)_n \tag{3.39}$$

which must satisfy the essential boundary condition. Then the functional has to be minimized with respect to the parameters C_n. In the present problem, for instance, we can use a power series expansion of the form

$$\bar{T}(r) = C_0 + \sum_{n=1}^{\infty} C_n r^n \tag{3.40}$$

For simplicity, we consider the case where $n = 2$, thus

$$\bar{T}(r) = C_0 + C_1 r + C_2 r^2 \tag{3.41}$$

Note that since $T(b) = 0$ so $C_0 = -C_1 b - C_2 b^2$, therefore,

$$\bar{T}(r) = C_1(r - b) + C_2(r^2 - b^2) \tag{3.42}$$

and

$$\frac{d\bar{T}(r)}{dr} = C_1 + 2C_2 r \tag{3.43}$$

By substituting Equations (3.42) and (3.43) into the functional (3.38) we obtain

$$\Pi(T) = k\pi \int_a^b (\frac{d\bar{T}}{dr})^2 r \, dr - 2\pi U a \bar{T} \tag{3.44}$$

then

$\Pi(T) =$

$$2\pi k \, [C_1 C_2] \begin{bmatrix} \frac{1}{2}(b^2 - a^2) & \frac{2}{3}(b^3 - a^3) \\ \frac{2}{3}(b^3 - a^3) & (b^4 - a^4) \end{bmatrix} \begin{bmatrix} C_1 \\ C_2 \end{bmatrix} - 2\pi U a [C_1 C_2] \begin{bmatrix} (a - b) \\ (a^2 - b^2) \end{bmatrix} \tag{3.45}$$

For Π to be a minimum, that is

$$\frac{\partial \Pi}{\partial C_i} = 0, \qquad r = 1, 2 \tag{3.46}$$

which provides two simultaneous equations as

$$2\pi k \begin{bmatrix} \frac{1}{2}(b^2 - a^2) & \frac{2}{3}(b^3 - a^3) \\ \frac{2}{3}(b^3 - a^3) & (b^4 - a^4) \end{bmatrix} \begin{bmatrix} C_1 \\ C_2 \end{bmatrix} = 2\pi U a \begin{bmatrix} (a - b) \\ (a^2 - b^2) \end{bmatrix} \tag{3.47}$$

If $a = 1$ and $b = 2$ then

$$\begin{bmatrix} \frac{3}{2} & \frac{14}{3} \\ \frac{14}{3} & 15 \end{bmatrix} \begin{bmatrix} C_1 \\ C_2 \end{bmatrix} = \frac{U}{k} \begin{bmatrix} -1 \\ -3 \end{bmatrix} \tag{3.48}$$

which may be solved to give

$$C_1 = -1.3846 \frac{U}{k}$$

$$C_2 = 0.2308 \frac{U}{k} \tag{3.49}$$

substituting Equation (3.49) into Equation (3.42) yields

$$\bar{T}(r) = -1.3846 \frac{U}{k}(r - 2) + 0.2308 \frac{U}{k}(r^2 - 4) \tag{3.50}$$

For the inner surface where $r = a = 1$, we have

$$\bar{T}(1) = 0.6922 \frac{U}{k} \tag{3.51}$$

whereas Equation (3.37) gives the analytical solution of $T(1) = 0.6931 \frac{U}{k}$. As we increase n we expect to converge to this solution [4].

3.3 Galerkin Weighted Residual Method

As mentioned in the previous section, for many continuum problems, a suitable variational principle is not available, since no corresponding functionals exist although their differential equations may well be formulated. As an alternative to solve such differential equations, we may use a variety of weighted residual methods. Let us consider the two dimensional problem (3.27) to (3.29). We begin by introducing the error, or *residual*, R_Ω in the approximation which is defined by

$$R_\Omega = \frac{\partial}{\partial x}(k\frac{\partial \bar{T}}{\partial x}) + \frac{\partial}{\partial y}(k\frac{\partial \bar{T}}{\partial y}) + Q \qquad (3.52)$$

where $\bar{T}$ contains trial functions and satisfies the Dirichlet boundary condition of $\bar{T} = T_0$ at Γ_T. The smaller the residual, the better the approximation. It should be noted that R_Ω is a function of position in Ω. Now, we attempt to reduce this residual as close to zero as possible. If we have

$$\int_\Omega W_i R_\Omega d\Omega = 0 \qquad i = 1, 2, ..., M \qquad (3.53)$$

where W_i is a set of arbitrary functions and $M \to \infty$, then it can be said that the R_Ω vanishes. We can adjust the free parameters C_i in $\bar{T}$ to approach this objective. Here W_i are called *weighting functions*. Expanding Equation (3.53), we have

$$\int_\Omega W_i \left[\frac{\partial}{\partial x}(k\frac{\partial \bar{T}}{\partial x}) + \frac{\partial}{\partial y}(k\frac{\partial \bar{T}}{\partial y}) + Q \right] dxdy = 0 \qquad (3.54)$$

A function $\bar{T}(x, y)$ that satisfies Equation (3.54) for every function W_i in Ω is a weak solution of the differential equation, whereas the strong solution $T(x, y)$ satisfies the differential equation at every point of Ω.

Similarly, we can treat the Neumann boundary condition as follows

$$R_{\Gamma_q} = k\frac{\partial \bar{T}}{\partial n} + \bar{q} \qquad (3.55)$$

where R_{Γ_q} is the residual on Γ_q. It is required that

$$\int_{\Gamma_q} \bar{W}_i R_\Gamma d\Gamma = 0 \qquad i = 1, 2, ..., M \qquad (3.56)$$

where $\bar{W}_i$ are the weighting functions on Γ_q. Thus,

$$\int_{\Gamma_q} \bar{W}_i \left[k\frac{\partial \bar{T}}{\partial n} + \bar{q} \right] d\Gamma = 0 \qquad (3.57)$$

Adding Equations (3.54) and (3.57), we have

$$\int_\Omega W_i \left[\frac{\partial}{\partial x}(k\frac{\partial \bar{T}}{\partial x}) + \frac{\partial}{\partial y}(k\frac{\partial \bar{T}}{\partial y}) + Q \right] dx dy + \int_{\Gamma_q} \bar{W}_i \left[k\frac{\partial \bar{T}}{\partial n} + \bar{q} \right] d\Gamma = 0$$

$$(3.58)$$

which may be rewritten, using Green's lemma, to give

$$- \int_\Omega \left[\frac{\partial W_i}{\partial x} k \frac{\partial \bar{T}}{\partial x} + \frac{\partial W_i}{\partial y} k \frac{\partial \bar{T}}{\partial y} \right] dx dy + \int_\Omega W_i Q dx dy +$$

$$\int_{\Gamma_T + \Gamma_q} k\frac{\partial \bar{T}}{\partial n} \bar{W}_i d\Gamma + \int_{\Gamma_q} \bar{W}_i \left[k\frac{\partial \bar{T}}{\partial n} + \bar{q} \right] d\Gamma = 0 \qquad (3.59)$$

Since both W_i and $\bar{W}_i$ are arbitrary, we can limit the choice of the weighting functions as

$$W_i = 0 \qquad \text{on} \quad \Gamma_T \qquad\qquad (3.60)$$

$$\bar{W}_i = -W_i \qquad \text{on} \quad \Gamma_q \qquad\qquad (3.61)$$

Now, it can be seen that the term involving the weighted integral of $\frac{\partial \bar{T}}{\partial n}$ on the boundary vanishes and the approximating equation becomes

$$\int_\Omega \left[\frac{\partial W_i}{\partial x} k \frac{\partial \bar{T}}{\partial x} + \frac{\partial W_i}{\partial y} k \frac{\partial \bar{T}}{\partial y} \right] dx dy - \int_\Omega W_i Q dx dy + \int_{\Gamma_q} W_i \bar{q} d\Gamma = 0 \quad (3.62)$$

This is the well known *weak form* of the steady state heat conduction equations. As for the variational method, the boundary condition

$$- k\frac{\partial T}{\partial n} = \bar{q} \qquad \text{on} \quad \Gamma_q \qquad\qquad (3.63)$$

is in some way natural for this problem as the formulation eliminates the need for an actual evaluation of $\frac{\partial T}{\partial n}$ on the boundaries and if $\bar{q} = 0$ such boundaries do not enter explicitly into Equation (3.62).

In order to utilize the weak form Equation (3.62) to obtain the approximation of the solution, one has to first choose appropriate trial functions, $N_i(x, y)$ to represent the real solution, that is

$$\bar{T}(x, y) = C_0 + \sum_{i=1}^{\infty} C_i N_i(x, y) \qquad\qquad (3.64)$$

which must satisfy the essential boundary conditions. The second task is to choose the weighting functions. The most popular weighted residual method is where the trial functions themselves are chosen as the weighting functions, thus

$$W_i(x, y) = N_i(x, y) \qquad\qquad (3.65)$$

This method was first used by Galerkin and is referred to the Galerkin method. Therefore, for Galerkin weighted residual method, the weak form of the 2-D heat conduction equations may be written as

$$\int_{\Omega} \left[\frac{\partial N_i}{\partial x} k \frac{\partial \bar{T}}{\partial x} + \frac{\partial N_i}{\partial y} k \frac{\partial \bar{T}}{\partial y} \right] dxdy - \int_{\Omega} N_i Q dxdy + \int_{\Gamma_q} N_i \bar{q} d\Gamma = 0 \quad (3.66)$$

Let us consider the same problem as given by Equations (3.35) and (3.36) in Section 2. The residual in Ω is

$$R_{\Omega} = \frac{1}{r} \frac{\partial}{\partial r} \left(rk \frac{\partial \bar{T}}{\partial r} \right) \quad (3.67)$$

and the residual on Γ_q is

$$R_{\Gamma_q} = k \frac{\partial \bar{T}}{\partial n} - U \quad (3.68)$$

The weak form in this case can be written as

$$\int_0^{2\pi} \int_a^b \left[\frac{\partial W_i}{\partial r} k \frac{\partial \bar{T}}{\partial r} \right] r dr d\theta - \int_0^{2\pi} W_i U d\theta = 0 \quad (3.69)$$

In order to obtain the solution, we use the two term approximation as in Equation (3.42) for T, that is

$$\bar{T}(r) = C_1(r - b) + C_2(r^2 - b^2) \quad (3.70)$$

which satisfies the essential boundary condition. According to the Galerkin method, the two weighting functions are

$$W_1 = r - b$$

and

$$W_2 = r^2 - b^2 \quad (3.71)$$

which vanish on the Γ_T. Substitution of Equations (3.70) and (3.71) into the weak form Equation (3.69) yields

$$2\pi k \begin{bmatrix} \frac{1}{2}(b^2 - a^2) & \frac{2}{3}(b^3 - a^3) \\ \frac{2}{3}(b^3 - a^3) & (b^4 - a^4) \end{bmatrix} \begin{bmatrix} C_1 \\ C_2 \end{bmatrix} = 2\pi U a \begin{bmatrix} (a - b) \\ (a^2 - b^2) \end{bmatrix} \quad (3.72)$$

which is identical to Equation (3.47). Using the same procedure the approximate solution can be obtained. Thus we see here that the Rayleigh-Ritz and Galerkin methods lead to identical results. For generality, we will use the Galerkin weighted residual approach to introduce the finite element method.

3.4   Finite Element Method in Two Dimensions

3.4.1   Introduction

In the previous sections we employed a single expression valid throughout the whole domain Ω to approximate the real solution, that is

$$\bar{T}(x,y) = C_0 + \sum_{i=1}^{\infty} C_i N_i(x,y) \qquad (3.73)$$

The integrals of the approximating equations, such as Equation (3.26) and Equation (3.62), were evaluated in one operation over this domain. An alternative approach is to divide the domain Ω into a number of nonoverlapping *subregions* or *finite elements*, Ω_e. The shape of the finite elements is generally restricted to simple polygons such as triangles and quadrilaterals in two dimensions, pyramids and triangular and rectangular prisms in three dimensions, and so on. The approximation $\bar{T}$ is constructed in a *piecewise* manner over such elements. In each element, the approximation solution assumes the form of a linear combination of prescribed functions, thus

$$\bar{T}(x,y) = \sum_{\Omega_1}^{i=1,P} C_i^{(1)} N_i^{(1)}(x,y) + \sum_{\Omega_2}^{i=1,P} C_i^{(2)} N_i^{(2)}(x,y) + .. + \sum_{\Omega_M}^{i=1,P} C_i^{(M)} N_i^{(M)}(x,y)$$

$$(3.74)$$

where P is the number of free parameters C_i^e in each element and M is the total number of elements. In such a case, the definite integrals occurring in the weak form Equation (3.66) can be obtained simply by summing the contributions from each element as

$$\sum_{e=1}^{M} \int_{\Omega_e} \left[\frac{\partial N_i^e}{\partial x} k \frac{\partial \bar{T}^e}{\partial x} + \frac{\partial N_i^e}{\partial y} k \frac{\partial \bar{T}^e}{\partial y} \right] dx dy - \sum_{e=1}^{M} \int_{\Omega_e} N_i^e Q dx dy$$

$$+ \sum_{e=1}^{M} \int_{\Gamma_{q_e}} N_i^e \bar{q} d\Gamma = 0 \qquad (3.75)$$

where

$$\bar{T}^e(x,y) = \sum_{\Omega_e}^{i=1,P} C_i^e N_i^e(x,y) \qquad (3.76)$$

and Ω_e is the area of an element and Γ_{q_e} denotes that portion of the boundary of Ω_e which lies on Γ_q. Summations involving Γ_{q_e} are therefore taken only over those elements Ω_e which lie immediately adjacent to the boundary.

The shapes of elements chosen are usually the same for a particular problem and the definition of trial functions over each element can be made in a repeatable manner. Therefore, we can work out the formulation in one element Ω_e and quite readily extend it to the whole domain Ω. It is here that the essential idea of the finite element method lies. In fact the example used in Sections 3.2 and 3.3 is a special case of the finite element method where a single element is used.

3.4.2 Conductivity Matrix and Load Vector

For each element, we have to evaluate the following integral

$$\int_{\Omega_e} \left[\frac{\partial N_i^e}{\partial x} k \frac{\partial \bar{T}^e}{\partial x} + \frac{\partial N_i^e}{\partial y} k \frac{\partial \bar{T}^e}{\partial y} \right] dxdy - \int_{\Omega_e} N_i^e Q dxdy + \int_{\Gamma_{qe}} N_i^e \bar{q} d\Gamma$$

$$(i = 1, P) \tag{3.77}$$

The third term of Equation (3.77) appears, if any boundary of this element Ω_e lies on the boundary Γ_q of Ω. Now, we assume that the $N_i^e(x, y)$ are analytically simple, such as low-order polynomials and the like and C_i^e are local or nodal values of the solution T, that is

$$\bar{T}^e(x, y) = \sum_{\Omega_e}^{j=1, P} T_j^e N_j^e(x, y) \tag{3.78}$$

Substituting Equation (3.78) into Equation (3.77), we obtain

$$\int_{\Omega_e} \left[\frac{\partial N_i^e}{\partial x} k \frac{\partial N_j^e}{\partial x} + \frac{\partial N_i^e}{\partial y} k \frac{\partial N_j^e}{\partial y} \right] T_j^e dxdy - \int_{\Omega_e} N_i^e Q dxdy + \int_{\Gamma_{qe}} N_i^e \bar{q} d\Gamma$$

$$(i, j = 1, P) \tag{3.79}$$

Equation (3.79) may be rewritten as

$$K_{ij}^e T_j^e - f_i^e \qquad (i, j = 1, P) \tag{3.80}$$

where

$$K_{ij}^e = \int_{\Omega_e} \left[\frac{\partial N_i^e}{\partial x} k \frac{\partial N_j^e}{\partial x} + \frac{\partial N_i^e}{\partial y} k \frac{\partial N_j^e}{\partial y} \right] dxdy \tag{3.81}$$

which is named as the element *conductivity matrix* and

$$f_i^e = \int_{\Omega_e} N_i^e Q dxdy - \int_{\Gamma_{qe}} N_i^e \bar{q} d\Gamma \qquad (i = 1, P) \tag{3.82}$$

which is called the element *load vector*. Therefore, the approximating equations can be written as

$$K_{ij} T_j = f_i \qquad (i, j = 1, NP) \tag{3.83}$$

where

$$K_{ij} = \sum_{e=1}^{M} K_{ij}^e \qquad (3.84)$$

$$f_i = \sum_{e=1}^{M} f_i^e \qquad (3.85)$$

and NP is the total number of unknowns, T_j. So, the global conductivity matrix K_{ij} and load vector f_i are assembled from each element. If the convective boundary condition (2.36) is involved, then the element conductivity matrix becomes

$$K_{ij}^e = \int_{\Omega_e} \left[\frac{\partial N_i^e}{\partial x} k \frac{\partial N_j^e}{\partial x} + \frac{\partial N_i^e}{\partial y} k \frac{\partial N_j^e}{\partial y} \right] dxdy + \int_{\Gamma_{q_e}} h N_i^e N_j^e d\Gamma \qquad (3.86)$$

and the element load vector becomes

$$f_i^e = \int_{\Omega_e} N_i^e Q dxdy + \int_{\Gamma_{q_e}} h T_f N_i^e d\Gamma \qquad (i = 1, P) \qquad (3.87)$$

For the general non-isotropic case, the element conductivity matrix (3.81) may be written in tensor notation as

$$K_{ij}^e = \int_{\Omega_e} \left[\frac{\partial N_i^e}{\partial x_l} k_{lm} \frac{\partial N_j^e}{\partial x_m} \right] d\Omega \qquad (3.88)$$

In order to discuss further details, the type of finite element to be used for analysis must be defined.

3.4.3   Triangular Elements

The simplest finite element geometry for the heat conduction problem in two dimensions is that obtained by covering Ω with triangles having nodes (or solution points) at the vertices (Figure 3.2) and in which the finite element solution $\bar{T}(x, y)$ consists of plane triangular "tiles". The triangular finite element in two spatial dimensions holds an important place, particularly for the purpose of adaptively remeshing, because it is very flexible in idealising an arbitrary domain Ω geometrically.

Here, the linear triangular element is used to demonstrate all essential features of two dimensional finite element analysis. The element has three nodes numbered anticlockwise as 1,2,3, or generally as i, j, m. A linear temperature variation is assumed throughout the element, each node has one degree of freedom corresponding to the free parameters, that is the unknown temperature value T_i^e at that point. Therefore the temperature at any coordinate position x, y within the element can be represented as

$$T^e = a_1 + a_2 x + a_3 y \qquad (3.89)$$

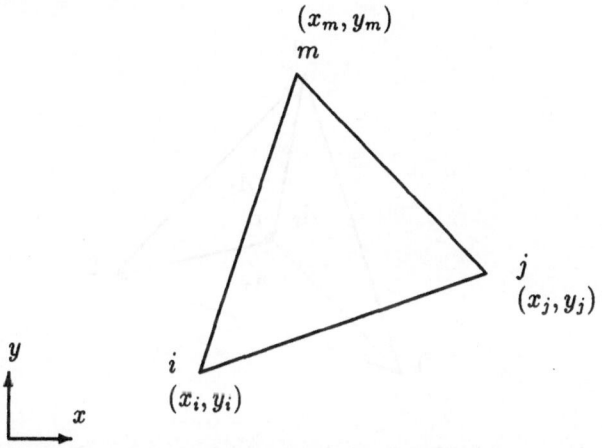

Figure 3.2: 3 node triangular element

since

$$T^e(x_1, y_1) = T_1^e$$
$$T^e(x_2, y_2) = T_2^e$$
$$T^e(x_3, y_3) = T_3^e \tag{3.90}$$

so, we have

$$T_1^e = a_1 + a_2 x_1 + a_3 y_1$$
$$T_2^e = a_1 + a_2 x_2 + a_3 y_2$$
$$T_3^e = a_1 + a_2 x_3 + a_3 y_3 \tag{3.91}$$

Solving Equation (3.91) for a_1, a_2 and a_3 yields

$$a_1 = \frac{1}{2A}[(x_2 y_3 - x_3 y_2)T_1^e + (x_3 y_1 - x_1 y_3)T_2^e + (x_1 y_2 - x_2 y_1)T_3^e]$$

$$a_2 = \frac{1}{2A}[(y_2 - y_3)T_1^e + (y_3 - y_1)T_2^e + (y_1 - y_2)T_3^e]$$

$$a_2 = \frac{1}{2A}[(x_3 - x_2)T_1^e + (x_1 - x_3)T_2^e + (x_2 - x_1)T_3^e] \tag{3.92}$$

where

$$A = \frac{1}{2} det \begin{bmatrix} 1 & x_1 & y_1 \\ 1 & x_2 & y_2 \\ 1 & x_3 & y_3 \end{bmatrix} = \quad \text{element area} \tag{3.93}$$

Therefore, we write

$$T^e = N_1^e T_1^e + N_2^e T_2^e + N_3^e T_3^e \tag{3.94}$$

where

$$N_1^e = \frac{1}{2A}(b_1 + c_1 x + d_1 y)$$

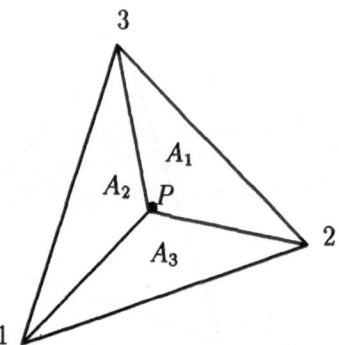

Figure 3.3: Area coordinate for triangular element

$$N_2^e = \frac{1}{2A}(b_2 + c_2 x + d_2 y)$$

$$N_3^e = \frac{1}{2A}(b_3 + c_3 x + d_3 y) \tag{3.95}$$

in which

$$b_i = x_j y_m - x_m y_i$$

$$c_i = y_j - y_m$$

$$d_i = x_m - x_j \tag{3.96}$$

and i, j and m are taken as $1, 2, 3$ in cyclic permutation. Here, the element trial functions N_i^e are named as the *shape functions*. It should be noted that $N_i^e(x, y)$ has a unit value at (x_i, y_i) of node i, whereas it has zero value at (x_j, y_j) of node j and (x_m, y_m) of node m. This is a characteristic of an element shape function.

Alternatively, for the linear three node triangle, the element shape functions can be simply defined by area coordinates. Considering Figure 3.3, the coordinates L_1, L_2 and L_3 are defined as

$$L_1 = \frac{A_1}{A}$$

$$L_2 = \frac{A_2}{A}$$

$$L_3 = \frac{A_3}{A} \tag{3.97}$$

where A is the area $\triangle 123$, A_1 the area $\triangle P23$, A_2 the area $\triangle P31$ and A_3 the area $\triangle P12$. Since P has only two Cartesian coordinates (x, y), it must be the case that the three coordinates (L_i) cannot be independent of each other. It is clear that the sum of the three area coordinates equals one, that is

$$L_1 + L_2 + L_3 = 1 \tag{3.98}$$

It is seen that L_i has a unit value at node i and zero value at other nodes with a linear variation in between. It is easily verified that the area coordinates L_1, L_2 and L_3 are identical to the linear shape functions N_1^e, N_2^e and N_3^e. The Cartesian coordinates (x, y) may also be expressed as

$$\begin{bmatrix} 1 \\ x \\ y \end{bmatrix} = \begin{bmatrix} 1 & 1 & 1 \\ x_1 & x_2 & x_3 \\ y_1 & y_2 & y_3 \end{bmatrix} \begin{bmatrix} L_1 \\ L_2 \\ L_3 \end{bmatrix} \tag{3.99}$$

The following formula is useful in evaluating integrals involved in the finite element procedure where triangular elements have been used, that is

$$\int_\Omega L_1^a L_2^b L_3^c d\Omega = \frac{a!b!c!}{(a+b+c+2)!} 2A \tag{3.100}$$

The conductivity matrix for an individual element is given by substituting Equation (3.95) into Equation (3.81), that is

$$K_{ij}^e = \int_{\Omega_e} \left[\frac{\partial N_i^e}{\partial x} k \frac{\partial N_j^e}{\partial x} + \frac{\partial N_i^e}{\partial y} k \frac{\partial N_j^e}{\partial y} \right] dx dy$$

$$= \int_{\Omega_e} \frac{k}{4A^2}[c_i c_j + d_i d_j] d\Omega = \frac{k}{4A}[c_i c_j + d_i d_j] \quad (i, j = 1, 2, 3) \tag{3.101}$$

and the complete element matrix may be written as

$$\mathbf{K}^e = \frac{k}{4A} \begin{bmatrix} c_1 c_1 + d_1 d_1 & c_1 c_2 + d_1 d_2 & c_1 c_3 + d_1 d_3 \\ c_2 c_1 + d_2 d_1 & c_2 c_2 + d_2 d_2 & c_2 c_3 + d_2 d_3 \\ c_3 c_1 + d_3 d_1 & c_3 c_2 + d_3 d_2 & c_3 c_3 + d_3 d_3 \end{bmatrix} \tag{3.102}$$

It can be seen $\mathbf{K}^e$ is symmetric. The element load vector contains two parts, that is

$$\mathbf{f}^e = \mathbf{f}_Q^e + \mathbf{f}_q^e \tag{3.103}$$

where

$$\mathbf{f}_Q^e = Q \int_{\Omega_e} \begin{bmatrix} N_i^e \\ N_j^e \\ N_m^e \end{bmatrix} d\Omega = \frac{QA}{3} \begin{bmatrix} 1 \\ 1 \\ 1 \end{bmatrix}$$

and

$$\mathbf{f}_q^e = -q \int_{\Gamma_q^e} \mathbf{N}^e d\Gamma \tag{3.104}$$

If side ij of the triangle is subjected to the natural boundary condition with a uniform flux q then

$$\mathbf{f}_q^e = \mathbf{f}_{q_{ij}}^e = -q \int_{\Gamma_{ij}^e} \begin{bmatrix} N_i^e \\ N_j^e \\ 0 \end{bmatrix} d\Gamma = \frac{-qL_{ij}}{2} \begin{bmatrix} 1 \\ 1 \\ 0 \end{bmatrix} \tag{3.105}$$

where L_{ij} is the length of element side ij.

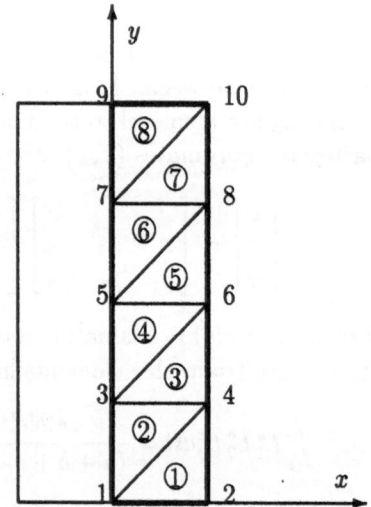

Figure 3.4: Finite element mesh

Now, we undertake a two-dimensional finite element analysis to demonstrate the process of assembly for the element matrices. The problem is steady state heat conduction in a rectangular region which we have analytically solved in the Section 3.1. The rectangular region is within $-1 \leq x \leq 1, 0 \leq y \leq 4$, The surfaces at $x = -1, x = 1$, and $y = 0$ are maintained at zero temperature, while the boundary surface at $y = 4$ is subjected to a prescribed temperature distribution $T_0 = 100$.

By taking advantage of the symmetry which exists in the problem, only half of the rectangle is idealised by 8 equal triangular three-noded elements with the nodal and element numbering shown in Figure 3.4.

The relationship between local and global numbering for each element is listed in the Table 3.1 in anticlockwise order.

From Equation (3.102), we obtain

$$\mathbf{K}^{(1)} = \mathbf{K}^{(2)} = ... = \mathbf{K}^{(8)} = \mathbf{K}^e = \frac{k}{2} \begin{bmatrix} 1 & -1 & 0 \\ -1 & 2 & -1 \\ 0 & -1 & 1 \end{bmatrix} \qquad (3.106)$$

where the thermal conductivity is assumed constant, say, $k = 2$. Since, $Q = 0$ and $q = 0$ in (3.103) and Equation (3.104),

$$\mathbf{f} = \sum_{e=1}^{M} \mathbf{f}^e = 0 \qquad (3.107)$$

According to (3.84), the global **K** matrix is assembled from the element

Table 3.1: Element Connectivities

No. element	i	j	m	No. element	i	j	m
1	1	2	4	5	5	6	8
2	4	3	1	6	8	7	5
3	3	4	6	7	7	8	10
4	6	5	3	8	10	9	7

matrices $\mathbf{K}^e$, thus

$$\mathbf{K} = \begin{bmatrix} K_{11} & K_{12} & K_{13} & K_{14} & & & & & & \\ & K_{22} & & K_{24} & & & & & & \\ & & K_{33} & K_{34} & & & & & & \\ & & & K_{44} & & K_{46} & & & & \\ & & & & K_{55} & K_{56} & K_{57} & K_{58} & & \\ & & & & & K_{66} & & K_{68} & & \\ & & SYM. & & & & K_{77} & K_{78} & K_{79} & K_{710} \\ & & & & & & & K_{88} & & K_{810} \\ & & & & & & & & K_{99} & K_{910} \\ & & & & & & & & & K_{1010} \end{bmatrix}$$

(3.108)

in which the non-zero terms are

$$K_{11} = K_{11}^{(1)} + K_{33}^{(2)} = 2 \qquad K_{12} = K_{12}^{(1)} = -1 \qquad K_{13} = K_{23}^{(2)} = -1$$
$$K_{14} = K_{13}^{(1)} + K_{31}^{(2)} = 0$$
$$K_{22} = K_{22}^{(1)} = 2 \qquad K_{24} = K_{32}^{(1)} = -1$$
$$K_{33} = K_{22}^{(2)} + K_{11}^{(3)} + K_{33}^{(4)} = 4 \qquad K_{34} = K_{21}^{(2)} + K_{12}^{(3)} = -2$$
$$K_{35} = K_{32}^{(4)} = -1 \qquad K_{36} = K_{13}^{(3)} + K_{31}^{(4)} = 0$$
$$K_{44} = K_{33}^{(1)} + K_{11}^{(2)} + K_{22}^{(3)} = 4 \qquad K_{46} = K_{23}^{(3)} = -1$$
$$K_{55} = K_{22}^{(4)} + K_{11}^{(5)} + K_{33}^{(6)} = 4 \qquad K_{56} = K_{21}^{(4)} + K_{12}^{(5)} = -2$$
$$K_{57} = K_{32}^{(6)} = -1 \qquad K_{58} = K_{13}^{(5)} + K_{31}^{(6)} = 0$$
$$K_{66} = K_{33}^{(3)} + K_{11}^{(4)} + K_{22}^{(5)} = 4 \qquad K_{68} = K_{23}^{(5)} = -1$$
$$K_{77} = K_{22}^{(6)} + K_{11}^{(7)} + K_{33}^{(8)} = 4 \qquad K_{78} = K_{21}^{(6)} + K_{12}^{(7)} = -2$$
$$K_{79} = K_{32}^{(8)} = -1 \qquad K_{7,10} = K_{13}^{(7)} + K_{31}^{(8)} = 0$$
$$K_{88} = K_{33}^{(5)} + K_{11}^{(6)} + K_{22}^{(7)} = 4 \qquad K_{8,10} = K_{23}^{(7)} = -1$$
$$K_{99} = K_{22}^{(8)} = 2 \qquad K_{9,10} = K_{21}^{(8)} = -1$$
$$K_{1010} = K_{33}^{(7)} + K_{11}^{(8)} = 2$$

It may be noted that the global $\mathbf{K}$ matrix is symmetric about the leading diagonal and that it is also banded. The bandwidth of the matrix is affected by the method of numbering of the nodes. In general, the smaller the bandwidth, the greater the computational efficiency of the solution process. Now, the set of equations may be written as

$$
\begin{bmatrix}
2 & -1 & -1 & 0 & 0 & 0 & 0 & 0 & 0 & 0 \\
-1 & 2 & 0 & -1 & 0 & 0 & 0 & 0 & 0 & 0 \\
-1 & 0 & 4 & -2 & -1 & 0 & 0 & 0 & 0 & 0 \\
0 & -1 & -2 & 4 & 0 & -1 & 0 & 0 & 0 & 0 \\
0 & 0 & -1 & 0 & 4 & -2 & -1 & 0 & 0 & 0 \\
0 & 0 & 0 & -1 & -2 & 4 & 0 & -1 & 0 & 0 \\
0 & 0 & 0 & 0 & -1 & 0 & 4 & -2 & -1 & 0 \\
0 & 0 & 0 & 0 & 0 & -1 & -2 & 4 & 0 & -1 \\
0 & 0 & 0 & 0 & 0 & 0 & -1 & 0 & 2 & -1 \\
0 & 0 & 0 & 0 & 0 & 0 & 0 & -1 & -1 & 2
\end{bmatrix}
\begin{bmatrix}
T_1 \\ T_2 \\ T_3 \\ T_4 \\ T_5 \\ T_6 \\ T_7 \\ T_8 \\ T_9 \\ T_{10}
\end{bmatrix}
=
\begin{bmatrix}
0 \\ 0 \\ 0 \\ 0 \\ 0 \\ 0 \\ 0 \\ 0 \\ 0 \\ 0
\end{bmatrix}
$$

$$(3.109)$$

From the boundary conditions we already know that

$$T_1 = T_2 = T_4 = T_6 = T_8 = 0$$

and,

$$T_9 = T_{10} = 100$$

Therefore, the equations corresponding to T_1, T_2, T_4, T_6 and T_8 are to be replaced by $T_1 = 0, T_2 = 0, T_4 = 0, T_6 = 0$ and $T_8 = 0$ and the equations corresponding to T_9 and T_{10} are to be replaced by $T_9 = 100$ and $T_{10} = 100$. Hence, the set of linear equations becomes

$$4T_3 - T_5 = 0$$

$$-T_3 + 4T_5 - T_7 = 0$$

$$-T_5 + 4T_7 - 100 = 0$$

or in matrix form as

$$
\begin{bmatrix}
4 & -1 & 0 \\
-1 & 4 & -1 \\
0 & -1 & 4
\end{bmatrix}
\begin{bmatrix}
T_3 \\ T_5 \\ T_7
\end{bmatrix}
=
\begin{bmatrix}
0 \\ 0 \\ 100
\end{bmatrix}
$$

We can use the Gaussian elimination method to reduce a linear equation system to an upper triangular form by successive elimination. However, in present case, the solution can be easily worked out as

$$T_3 = 1.78570$$

$$T_5 = 7.14286$$

$$T_7 = 26.7857$$

whereas the analytical solution gives

$$T_3^* = 1.0940$$

$$T_5^* = 5.4880$$

$$T_7^* = 26.094$$

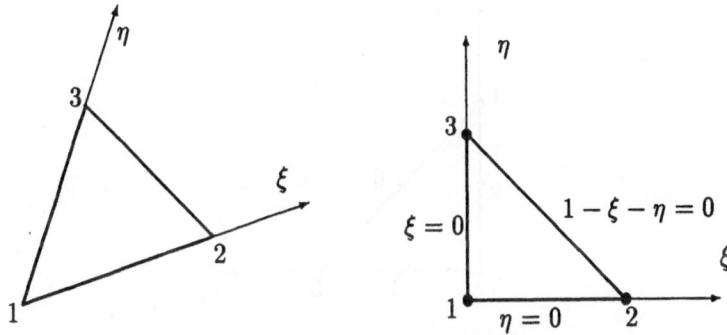

Figure 3.5: Mapping from general triangle to right-isosceles triangle

The finite element results approach the analytical solution when the mesh is further refined.

3.4.4 Natural Coordinate System

There is another way to create shape functions for finite elements. As we know, a simple element, such as the square or the right-isosceles triangle can be mapped into a more complex shape in the global coordinate system.

Here, we consider the triangular mapping first. In a typical triangle (Figure 3.5), a skew coordinate system ξ, η is constructed by its side 1-2 and 1-3. The coordinates ξ, η at nodes 1, 2 and 3 are $(0,0)$, $(1,0)$ and $(0,1)$ respectively. Therefore the actual element is transformed into a right-isosceles triangle in ξ, η which is called the master element. The functions defining three sides of the master element are given by $\eta = 0, 1 - \xi - \eta = 0$ and $\xi = 0$. Now, we can easily find out the shape functions as follows

$$N_1^e = 1 - \xi - \eta \tag{3.110}$$

which has a unit value at node 1 and zero value at nodes 2, 3.

$$N_2^e = \xi \tag{3.111}$$

which has a unit value at node 2 and zero value at nodes 3, 1.

$$N_3^e = \eta \tag{3.112}$$

which has a unit value at node 3 and zero value at nodes 1, 2

The global coordinate (x,y) can be expressed as

$$x = N_1^e x_1 + N_2^e x_2 + N_3^e x_3$$

$$y = N_1^e y_1 + N_2^e y_2 + N_3^e y_3 \tag{3.113}$$

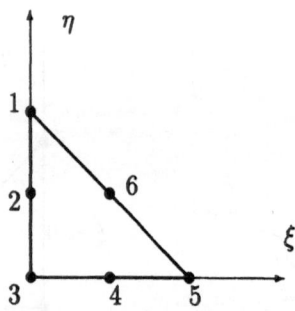

Figure 3.6: 6 node triangular element

which is identical to Equation (3.99). In fact [5], here ξ is the same as L_2 and η is the same as L_3 and

$$N_1^e = 1 - \xi - \eta = L_1 \tag{3.114}$$

In the derivation of the element conductivity matrix (3.81), it has been necessary to establish the shape function derivatives with respect to the (x,y) coordinates. Since by chain rule, we have

$$\frac{\partial N_i}{\partial \xi} = \frac{\partial N_i}{\partial x}\frac{\partial x}{\partial \xi} + \frac{\partial N_i}{\partial y}\frac{\partial y}{\partial \xi}$$

$$\frac{\partial N_i}{\partial \eta} = \frac{\partial N_i}{\partial x}\frac{\partial x}{\partial \eta} + \frac{\partial N_i}{\partial y}\frac{\partial y}{\partial \eta} \tag{3.115}$$

Therefore, the required derivatives $\frac{\partial N_i}{\partial x}$ and $\frac{\partial N_i}{\partial y}$ can be obtained by inversion as

$$\begin{bmatrix} \frac{\partial N_i}{\partial x} \\ \frac{\partial N_i}{\partial y} \end{bmatrix} = \mathbf{J}^{-1} \begin{bmatrix} \frac{\partial N_i}{\partial \xi} \\ \frac{\partial N_i}{\partial \eta} \end{bmatrix} \tag{3.116}$$

where $\mathbf{J}$ is called the *Jacobian matrix*, given by

$$\mathbf{J} = \begin{bmatrix} \frac{\partial x}{\partial \xi} & \frac{\partial y}{\partial \xi} \\ \frac{\partial x}{\partial \eta} & \frac{\partial y}{\partial \eta} \end{bmatrix} \tag{3.117}$$

This mapping technique is specially useful for high order elements. For instance, the six node triangular element shown in Figure 3.6 has shape functions as

$$\begin{aligned}
N_1^e &= \eta(2\eta - 1) \\
N_2^e &= 4\eta(1 - \xi - \eta) \\
N_3^e &= (1 - \xi - \eta)(1 - 2\xi - 2\eta) \\
N_4^e &= 4\xi(1 - \xi - \eta) \\
N_5^e &= \xi(2\xi - 1) \\
N_6^e &= 4\xi\eta
\end{aligned} \tag{3.118}$$

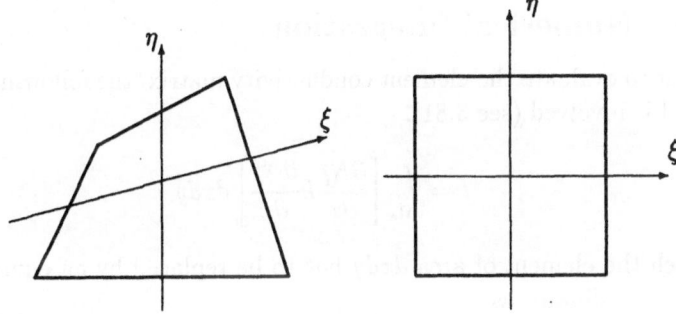

Figure 3.7: Mapping from quadrilateral to square

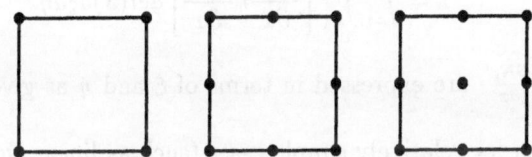

Figure 3.8: 4,8 and 9 node elements

which are quadratic in (ξ, η).

In addition to triangles, quadrilateral elements are widely used, Figure 3.7 shows the mapping from a general quadrilateral to a square. The shape functions of some typical quadrilateral elements (Figure 3.8) are listed as follows.

Bilinear 4-node element

$$N_i^e = (1 + \xi\xi_i)(1 + \eta\eta_i) \tag{3.119}$$

Serendipity 8-node element

$$\text{Corner nodes: } N_i^e = (1 + \xi\xi_i)(1 + \eta\eta_i)(\xi\xi_i + \eta\eta_i - 1)/4$$
$$\text{Midside nodes: } N_i^e = \xi_i^2(1 + \xi\xi_i)(1 - \eta^2)/2 + \eta_i^2(1 + \eta\eta_i)(1 - \xi^2)/2$$

$$\tag{3.120}$$

Lagrangian 9-node element

$$N_i^e = [\xi\xi_i(1 + \xi\xi_i)/2 + (1 - \xi^2)(1 - \xi_i^2)][\eta\eta_i(1 + \eta\eta_i)/2 + (1 - \eta\eta^2(1 - \eta_i^2]$$
$$\tag{3.121}$$

where ξ_i, η_i are the natural coordinates at the element node i.

3.4.5 Numerical Integration

In order to evaluate the element conductivity matrix, the following typical integral is involved (see 3.81).

$$I = \int_{\Omega_e} \left[\frac{\partial N_i^e}{\partial x} k \frac{\partial N_j^e}{\partial x} \right] dx dy \qquad (3.122)$$

in which the element of area $dx dy$ has to be replaced by an equivalent in the ξ, η coordinates as

$$dx dy = det(\mathbf{J}) d\xi d\eta \qquad (3.123)$$

Therefore Equation (3.123) can be rewritten in terms of integration over a square domain as

$$I = \int_{-1}^{1} \int_{-1}^{1} \left[\frac{\partial N_i^e}{\partial x} k \frac{\partial N_j^e}{\partial x} \right] det(\mathbf{J}) d\xi d\eta \qquad (3.124)$$

where $\frac{\partial N_i^e}{\partial x}, \frac{\partial N_j^e}{\partial x}$ are expressed in terms of ξ and η as given by Equation (3.116).

Other than for relatively simple cases (such as linear elements in triangles or bilinear elements in rectangles) the integrals involved in Equation (3.124) cannot be expressed in simple analytic form. Therefore, numerical integration has to be used to evaluate such integrals. Here, Gauss-Legendre quadrature rules are adopted.

Numerical quadrature formulas in triangular elements have the form

$$I = \int_{\Omega} f(\xi, \eta) d\xi d\eta = \sum_{i=1}^{m} a_i f(\xi_i, \eta_i) \qquad (3.125)$$

where ξ_i, η_i are natural coordinates of the sampling points, a_i are weighting factors and m is the total number of integration points. The positions of the sampling points and the values of the weighting factors for $m = 1, 3, 4, 7$ are listed in Table 3.2 in which p is the degree of polynomial in ξ or η which can be integrated exactly. The typical 4-point integration rule is shown in Figure 3.9.

Numerical quadrature formulas in quadrilateral elements have the form

$$I = \int_{-1}^{1} \int_{-1}^{1} f(\xi, \eta) d\xi d\eta = \sum_{i=1}^{n} \sum_{j=1}^{n} a_i a_j f(\xi_i, \eta_j) \qquad (3.126)$$

where ξ_i, η_i are natural coordinates of the sampling points, a_i, a_j are weighting factors and n is number of integration points in one direction. The positions of the sampling points and the values of the weighting factors for $n = 1, 2, 3$ are listed in Table 3.3 in which p is the degree of polynomial in ξ or η which can be integrated exactly. A 2×2 integration rule is shown in Figure 3.10

Table 3.2: Quadrature points and weighting factors for triangles

m	p	ξ_i	η_j	a_i
1	1	$\frac{1}{3}$	$\frac{1}{3}$	1
3	2	$\frac{1}{2}$	$\frac{1}{2}$	
		0	$\frac{1}{2}$	$\frac{1}{3}$
		$\frac{1}{2}$	0	
4	3	$\frac{1}{3}$	$\frac{1}{3}$	$-\frac{27}{48}$
		$\frac{3}{5}$	$\frac{1}{5}$	
		$\frac{1}{5}$	$\frac{3}{5}$	$\frac{25}{48}$
		$\frac{1}{5}$	$\frac{1}{5}$	
7	4	$\frac{1}{3}$	$\frac{1}{3}$	0.2250000000
		0.0597158717	0.4701420641	
		0.4701420641	0.0597158717	0.1323941527
		0.4701420641	0.4701420641	
		0.7974269853	0.7974269853	
		0.1012865073	0.7974269853	0.1259391805
		0.1012865073	0.1012865073	

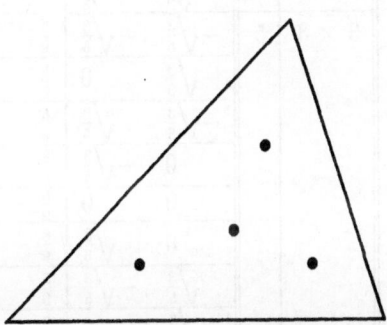

Figure 3.9: 4 point Gauss rule for triangles

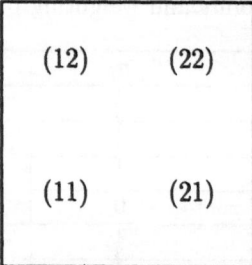

Figure 3.10: 2×2 point Gauss rule for quadrilaterals

Table 3.3: Quadrature points and weighting factors for quadrilaterals

$n \times n$	p	ξ_i	η_j	a_i	a_j
1×1	1	0	0	2	2
2×2	3	$-\frac{1}{\sqrt{3}}$	$-\frac{1}{\sqrt{3}}$		
		$-\frac{1}{\sqrt{3}}$	$\frac{1}{\sqrt{3}}$	1	1
		$\frac{1}{\sqrt{3}}$	$-\frac{1}{\sqrt{3}}$		
		$\frac{1}{\sqrt{3}}$	$\frac{1}{\sqrt{3}}$		
3×3	5	$-\sqrt{\frac{3}{5}}$	$-\sqrt{\frac{3}{5}}$	$\frac{5}{9}$	$\frac{5}{9}$
		$-\sqrt{\frac{3}{5}}$	0	$\frac{5}{9}$	$\frac{8}{9}$
		$-\sqrt{\frac{3}{5}}$	$\sqrt{\frac{3}{5}}$	$\frac{5}{9}$	$\frac{5}{9}$
		0	$-\sqrt{\frac{3}{5}}$	$\frac{8}{9}$	$\frac{5}{9}$
		0	0	$\frac{8}{9}$	$\frac{8}{9}$
		0	$\sqrt{\frac{3}{5}}$	$\frac{8}{9}$	$\frac{5}{9}$
		$\sqrt{\frac{3}{5}}$	$-\sqrt{\frac{3}{5}}$	$\frac{5}{9}$	$\frac{5}{9}$
		$\sqrt{\frac{3}{5}}$	0	$\frac{5}{9}$	$\frac{8}{9}$
		$\sqrt{\frac{3}{5}}$	$\sqrt{\frac{3}{5}}$	$\frac{5}{9}$	$\frac{5}{9}$

References

[1] M.N.Ozisik. *Basic Heat Transfer.* McGraw-Hill Book Company, New York, U.S.A., 1977.

[2] O.C.Zienkiewicz and K.Morgan. *Finite Elements and Approximation.* John Wiley and Sons, New York, U.S.A., 1983.

[3] H.S.Carslaw and J.C.Jaeger. *Conduction of Heat in Solids.* Clarendon Press, Oxford, 1959.

[4] D.R.J.Owen and E.Hinton. *A Simple Guide to Finite Elements.* Pineridge Press, Swansea, U.K., 1980.

[5] G.Carey and J.T.Oden. *Finite Elements, Vol II.* Pineridge Press, Englewood Cliffs, New Jersey 07632, 1983.

Chapter 4

Temporal Discretisation for Heat Conduction

4.1 Introduction

Problems where the temperature field at various points in the domain varies with time are referred to as *transient* problems, as opposed to *steady state* problems, where the temperature remains constant at a given point in the domain, for all times. In structural mechanics transient problems are analogous with dynamics and steady state problems are analogous with statics. The finite element discretisation discussed in the previous chapter was limited to the heat conduction equations without the term containing the temporal derivative. Although, most real life heat transfer problems are time-dependent, for many engineering problems it is sufficient to calculate a steady spatial temperature field. For example, the temperature field for electrical or mechanical machinery in operational conditions may be calculated as a steady state problem governed by the steady heat conduction equation and appropriate boundary conditions, using the procedure outlined in the previous chapter. However, there are other problems where the transient effects cannot be ignored. For example, it may be required to calculate the temperature field for machinery subjected to time-dependent or cyclic thermal loading. Other examples are phase change problems (solidification, melting etc.). For such problems the complete heat conduction equations including the temporal derivative term must be used. Therefore, a temporal discretisation of the transient heat conduction equations is required in addition to the spatial discretisation.

In the following sections we will present the additional techniques required for the finite element analysis of transient conduction problems. Apart from the basic temporal discretisation methods we will discuss other relevant topics such as stability and automatic timestep selection. To illustrate the transient solution method, a numerical test is conducted on a benchmark problem at the end of the chapter.

4.2 Finite Element Discretisation of the Transient Conduction Equation

It was shown in the previous chapter that the Galerkin form of the weighted residual method when applied to the steady heat conduction equation, gives identical results to those achieved by Rayleigh-Ritz approximation of the variational functional of Equation (3.30). We will therefore directly use the Galerkin weighted residual method to derive the finite element form of the transient equation. The residual for the transient heat conduction equation is defined as:

$$R_\Omega = \frac{\partial}{\partial x}(k\frac{\partial \bar{T}}{\partial x}) + \frac{\partial}{\partial y}(k\frac{\partial \bar{T}}{\partial y}) + F - \rho c\frac{\partial \bar{T}}{\partial t} \tag{4.1}$$

The subsequent steps of the derivation to obtain the Galerkin form of the weighted residual expression remain the same as outlined in the previous chapter, *i.e.* Equation (3.53) to Equation (3.66). The only exception is the presence of the temporal derivative term in the weak form of the transient conduction equation in addition to the terms in Equation (3.66). The Galerkin weak form of the transient heat conduction equation may therefore be written as,

$$\int_\Omega \rho c N_i \frac{\partial \bar{T}}{\partial t} dx dy + \int_\Omega \left[\frac{\partial N_i}{\partial x}k\frac{\partial \bar{T}}{\partial x} + \frac{\partial N_i}{\partial y}k\frac{\partial \bar{T}}{\partial y} \right] dx dy$$

$$- \int_\Omega N_i Q dx dy + \int_{\Gamma_q} N_i \bar{q} d\Gamma = 0 \tag{4.2}$$

To obtain the algebraic system of equations from the weak form of Equation (4.2) we use the familiar strategy of dividing the solution domain into subregions or *elements*, and use the element shape functions as discussed earlier to define the solution within the element, based on the nodal values. In this case however, we apply these shape functions to the temporal derivative, *i.e.*,

$$\frac{\partial \bar{T}^e}{\partial t}(x,y) = \sum_{\Omega_e}^{j=1,P} \frac{\partial T_j^e}{\partial t} N_j^e(x,y) \tag{4.3}$$

It may be noted that the weighting functions in Equation (4.3) are identical to the trial or shape functions in Equation (4.3) above. This will result in the final Galerkin finite element form of the discretised equations, which may be written as,

$$\sum_{e=1}^{M} \int_{\Omega_e} \rho c N_i^e N_j^e \frac{\partial T_j^e}{\partial t} dx dy + \sum_{e=1}^{M} \int_{\Omega_e} \left[\frac{\partial N_i^e}{\partial x}k\frac{\partial N_j^e}{\partial x} + \frac{\partial N_i^e}{\partial y}k\frac{\partial N_j^e}{\partial y} \right] T_j^e dx dy$$

$$- \sum_{e=1}^{M} \int_{\Omega_e} N_i^e Q dx dy + \sum_{e=1}^{M} \int_{\Gamma_{qe}} N_i^e \bar{q} d\Gamma = 0 \qquad (i,j = 1,P) \tag{4.4}$$

Here M is the total number of elements and P is the number of nodes in each element. The first term of the above equation represents the global *heat capacity* matrix $\mathbf{C}$.

Equation (4.4) represents the spatially discretised finite element form of the transient heat conduction equation (2.22). It is a first order system of ordinary differential equations with respect to time. To arrive at the final set of algebraic equations which may be solved using the digital computer we must discretise the time domain. Again, there are several ways of accomplishing the discretisation of the time domain. In the finite difference discretisation the time domain is divided into many increments and the equation is solved step by step in the given space domain. Alternatively, an algorithm named 'space-time' finite element method [1], where both the space and time domains are discretised by piecewise functions. Recently, another algorithm named 'space-time semi-analytic' method has been developed [2]. In this method a series representation is taken in the time domain, then the convolution-type variational principle has to be employed. However, due to the conceptual simplicity of the time dimension, simpler finite difference approximations are generally favoured. Most schemes currently used are constructed in this way.

In this chapter we will present the most commonly used methods for integrating the spatially discretised heat conduction equations in time. This first order system of equations may be written as:

$$\mathbf{C}\dot{\mathbf{T}} + \mathbf{K}\mathbf{T} = \mathbf{F} \tag{4.5}$$

Here, $\mathbf{K}$ represents the global conductivity matrix which is obtained from the summation of the element conductivity matrices as in Equation (3.81). Also, $\mathbf{F}$ represents the global load vector obtained from assembling the element load vectors as in Equation (3.82).

The objective of a time integration scheme is to find the unknown values $\mathbf{T}_{n+1}$ at time point t_{n+1} from the known information. The values $\mathbf{T}_n$ are known at time point t_n and so is $\mathbf{F}$ in the small time interval Δt. The same scheme is continued until the time of interest is reached, therefore these schemes are referred to as recurrence relations. The time intervals Δt may be considered as elements of time and appropriate shape functions may be defined at the ends of the interval, *i.e.* t_{n+1} and t_n, to establish the variation of the temperature field within the interval.

4.3 Recurrence Relations

The methods used for time discretisation of the heat equation may be broadly classified as single step and multistep methods. Of these the single step methods are the most popular because of their simplicity and flexibility. The main operational advantage of these methods is that the timestep size may be varied during the analysis. We will present in the following pages the details of a commonly used method for the equation system (4.5), the *Generalised mid-point* or *trapezoidal* family of methods, see references [3, Chapter 8] and [4, Chapter 2] for details.

4.3.1   Generalised Trapezoidal and Mid-point Family of Methods

The trapezoidal method when applied to Equation (4.5), can be written as follows,

$$\mathbf{C}(\mathbf{T}_{n+1}, t_{n+1})\dot{\mathbf{T}}_{n+1} + \mathbf{K}(\mathbf{T}_{n+1}, t_{n+1})\mathbf{T}_{n+1} = \mathbf{F}_{n+1} \qquad (4.6)$$

and

$$\frac{\dot{\mathbf{T}}_{n+1} + \dot{\mathbf{T}}_n}{2} = \frac{\mathbf{T}_{n+1} - \mathbf{T}_n}{\Delta t_n} \qquad (4.7)$$

subscripts n represent the nth time step. If not indicated otherwise Δt will mean Δt_n. Substituting Equation (4.7) in Equation (4.6), we obtain,

$$\left[\frac{2\mathbf{C}_{n+1}}{\Delta t} + \mathbf{K}_{n+1}\right](\mathbf{T}_{n+1}) = [\mathbf{C}_{n+1}]\left(\frac{2}{\Delta t}\mathbf{T}_n + \dot{\mathbf{T}}_n\right) + (\mathbf{F}_{n+1}) \qquad (4.8)$$

This method involves the calculation of the derivatives on the right hand side. Here, $\mathbf{C}_{n+1}$ etc. mean $\mathbf{C}(\mathbf{T}_{n+1}, t_{n+1})$.

The generalised mid-point family of methods [4, page 145] is written as,

$$\mathbf{C}(\mathbf{T}_{n+\alpha}, t_{n+\alpha})\dot{\mathbf{T}}_{n+\alpha} + \mathbf{K}(\mathbf{T}_{n+\alpha}, t_{n+\alpha})\mathbf{T}_{n+\alpha} = \mathbf{F}(\mathbf{T}_{n+\alpha}, t_{n+\alpha}) \qquad (4.9)$$

where,

$$
\begin{aligned}
\mathbf{T}_{n+\alpha} &= (1-\alpha)\mathbf{T}_n + \alpha\mathbf{T}_{n+1} \\
\dot{\mathbf{T}}_{n+\alpha} &= \frac{\mathbf{T}_{n+1} - \mathbf{T}_n}{\Delta t} \\
t_{n+\alpha} &= t_n + \alpha\Delta t
\end{aligned}
\qquad (4.10)
$$

Substituting Equation (4.10) into Equation (4.9) we obtain,

$$\left[\frac{\mathbf{C}_{n+\alpha}}{\Delta t} + \alpha\mathbf{K}_{n+\alpha}\right](\mathbf{T}_{n+1}) = \left[\frac{\mathbf{C}_{n+\alpha}}{\Delta t} - (1-\alpha)\mathbf{K}_{n+\alpha}\right](\mathbf{T}_n) + (\mathbf{F}_{n+\alpha})$$

$$(4.11)$$

No calculation of derivatives is necessary for this method. By changing the values of α from 0 to 1, different members of this family of methods are identified, *i.e.*,

$\alpha = 0$ -Forward Difference or Forward Euler.

$\alpha = \frac{1}{2}$ -Midpoint rule or Crank Nicolson.

$\alpha = \frac{2}{3}$ -Galerkin.

$\alpha = 1$ -Backward Difference or Backward Euler.

All, except the first (forward Euler), of the above schemes are implicit, i.e., they require matrix inversion for solution.

4.3.2 Convergence

Transient algorithms must have the property of convergence to be viable. This means that the time discretisation error must approach zero as the size of the time interval Δt approaches zero. According to the Lax equivalence theorem, if the algorithm is stable and consistent it will converge.

To analyse the stability properties of the generalised mid-point algorithms we consider the homogeneous version of Equation (4.5),

$$\mathbf{C}\dot{\mathbf{T}} + \mathbf{K}\mathbf{T} = 0 \tag{4.12}$$

which is then written in a SDOF (single degree of freedom) form. This can be done by a modal decomposition procedure. We begin by substituting,

$$\mathbf{T} = e^{-\lambda t}\mathbf{\Phi}$$

therefore

$$\dot{\mathbf{T}} = -\lambda e^{-\lambda t}\mathbf{\Phi}$$

which leads to,

$$(\mathbf{K} - \lambda\mathbf{C})\,\mathbf{\Phi} = 0 \tag{4.13}$$

Equation (4.13) represents an eigenvalue problem with the associated properties, *i.e.*,

$$\mathbf{\Phi}^T\mathbf{C}\mathbf{\Phi} = \mathbf{I} \tag{4.14}$$

and

$$\mathbf{\Phi}^T\mathbf{K}\mathbf{\Phi} = \lambda\mathbf{I} \tag{4.15}$$

where, $\mathbf{I}$ is the identity matrix, λ and $\mathbf{\Phi}$ are the eigenvalues and the eigenvectors, so that ($\mathbf{C}$ and $\mathbf{K}$ being positive definite),

$$0 \leq \lambda_1 \leq \lambda_2 \leq \dots \leq \lambda_N$$

where N is the total number of equations. We can further write,

$$\mathbf{T} = \sum_{i=1}^{N} T_i\mathbf{\Phi}_i \tag{4.16}$$

and

$$\dot{\mathbf{T}} = \sum_{i=1}^{N} \dot{T}_i\mathbf{\Phi}_i \tag{4.17}$$

where T_i and $\dot{T}_i$ are Fourier coefficients. Substituting Equations (4.16) and (4.17) in Equation (4.12) and premultiplying by $\mathbf{\Phi}^T$ we obtain,

$$\sum_{i=1}^{N} \left(\dot{T}_i(\mathbf{\Phi}^T\mathbf{C}\mathbf{\Phi}) + T_i(\mathbf{\Phi}^T\mathbf{K}\mathbf{\Phi})\right) = 0 \tag{4.18}$$

substuting Equations (4.14) and (4.15) in Equation (4.18) gives us N decoupled SDOF equations of the form,

$$\dot{T} + \lambda T = 0 \tag{4.19}$$

This is a first order ordinary differential equation which may be exactly solved between the general interval t_n and t_{n+1}. This gives us the solution,

$$T_{n+1} = e^{-\lambda(t_{n+1}-t_n)}T_n \tag{4.20}$$

as $(t_{n+1} - t_n) > 0$ the above solution implies that $|T_{n+1}| < |T_n|$. If we use the generalised midpoint rule to discretise Equation (4.19), we obtain,

$$(1 + \alpha\Delta t\lambda)T_{n+1} = (1 - (1-\alpha)\Delta t\lambda)T_n \tag{4.21}$$

or

$$T_{n+1} = \frac{(1 - (1-\alpha)\Delta t\lambda)}{(1 + \alpha\Delta t\lambda)}T_n = AT_n \tag{4.22}$$

where, A is defined as the *amplification factor*. Therefore for stability we have the condition $|A| < 1$, or

$$-1 < \frac{(1 - (1-\alpha)\Delta t\lambda)}{(1 + \alpha\Delta t\lambda)} < 1 \tag{4.23}$$

Solving the right hand inequality we obtain $1 > 0$, which means it is satisfied for all values of α. The left hand inequality however, yields the following equation for values of $\alpha < \frac{1}{2}$,

$$\Delta t < \frac{2}{(1 - 2\alpha)\lambda} \tag{4.24}$$

which implies conditional stability for these values of α, and unconditional stability for all other values. Of the methods mentioned in the previous section the Forward Euler method is the only one with conditional stability. Another interesting aspect of this analysis is the *oscillation limit* at $A = 0$, *i.e.*

$$\Delta t = \frac{1}{(1 - \alpha)\lambda} \tag{4.25}$$

For the unconditionally stable methods this limit is lowest for $\alpha = \frac{1}{2}$. This is because, for large values of $\lambda\Delta t$, the value of $A \simeq -1$ and the high modal components behave like $(-1)^n$, which manifests itself in the solution as oscillations. This effect may be filtered out by a simple averaging technique at each step [5].

The above stability limits may be made more meaningful if they are written in terms of thermal properties instead of eigenvalues. If we consider a single element of length h with one end at a fixed temperature of zero (at $x = 0$). The temperature in the rest of the element may then be defined by means of a single shape function $\frac{x}{h}$ multiplied by the temperature at the free end. If this shape function is used for discretising the homogeneous equation 4.12, we obtain the eigenvalue as,

$$\lambda = \frac{3k}{h^2 \rho c} \tag{4.26}$$

which for the explicit Forward Euler method ($\alpha = 0$) gives,

$$\Delta t < \frac{2}{3} \frac{\rho c}{k} h^2 \tag{4.27}$$

this limit for lumped (instead of consistent mass or capacitance matrices) is,

$$\Delta t < \frac{\rho c}{k} h^2 \tag{4.28}$$

This result may be applied to general 2-D or 3-D problems if the minimum element size (h_{min}) in the mesh is used which corresponds to the maximum eigenvalue. It is seen from these results that the critical time step for explicit schemes (at which instability impends) depends upon the square of the element size. In practice, this limit can be prohibitively small, especially when higher order elements are used (where h corresponds to the distance between two nearest nodes). For this reason, implicit methods are generally preferred for heat conduction problems.

Algorithms based on rational assumptions are consistent, where consistency is defined as follows. Consider Equation (4.22), which may be written as under with a load term L_n,

$$T_{n+1} - AT_n - L_n = 0 \tag{4.29}$$

replacing T_n and T_{n+1} above with corresponding exact values we have,

$$T(t_{n+1}) - AT(t_n) - L_n = \Delta t \tau(t_n) \tag{4.30}$$

where $\tau(t_n)$ is called the local truncation error. For consistency we must have

$$\tau(t) \leq C \Delta t^r \qquad \text{for all t} \tag{4.31}$$

where C is a constant and $r > 0$, r being the rate of convergence. The generalised midpoint and trapezoidal families of methods are consistent with $r = 1$ for all values of α except for $\alpha = \frac{1}{2}$, for which $r = 2$. For the proof of this the reader may refer to [3].

The above stability analysis may be extended to the trapezoidal rule as well, using the general equivalence of the trapezoidal and the mid-point family [6], as suggested by Hughes [4, page 150]. As far as accuracy is concerned the mid-point rule is to be preferred [7]. Also, Cliffe [8] has shown that the generalised mid-point rule conserves linear and quadratic quantities, while the trapezoidal rule conserves only the linear ones.

4.4   Automatic Time Step Selection

It is advantageous in most transient problems to adapt the time step-size to the temporal gradients of solution to reduce running costs. This may be done by simply reducing or increasing the time step-size depending upon the number of iterations required for convergence in the previous step, however this is only an ad-hoc rule. Gresho et. al. [9] have used a method based on the difference of the local time discretisation error between a predictor *i.e.*, AB (Adams- Bashforth) formula and a corrector TR (Trapezoidal rule). A description of the method is given below.

The AB formula gives for an equation, $y' = f$, (where y' represents the first derivative of y with respect to time, and so on).

$$y_{n+1}^p = y_n + \frac{\Delta t_n}{2} \left[\left(2 + \frac{\Delta t_n}{\Delta t_{n-1}} \right) y_n' - \left(\frac{\Delta t_n}{\Delta t_{n-1}} \right) y_{n-1}' \right] \qquad (4.32)$$

where p stands for predictor and n is the time level.

For the local discretisation error we write the exact solution at time t_{n+1} using the Taylor series expansion,

$$y(t_{n+1}) = y_n + \Delta t_n y_n' + \frac{\Delta t_n^2}{2} y_n'' + \frac{\Delta t_n^3}{6} y_n''' - O(\Delta t_n)^4 \qquad (4.33)$$

where the last term represents the error due to non-inclusion of higher order Taylor series terms. Subtracting Equation (4.33) from Equation (4.32) and simplifying, we get,

$$y_{n+1}^p - y(t_{n+1}) = -\frac{1}{12} \left(2 + 3 \frac{\Delta t_{n-1}}{\Delta t_n} \right) \Delta t_n^3 y_n''' + O(\Delta t_n)^4 \qquad (4.34)$$

We have from the trapezoidal rule,

$$y_{n+1} = y_n + \frac{\Delta t_n}{2} \left(f_{n+1} + f_n \right)$$

$$= y_n + \frac{\Delta t_n}{2} \left(y_{n+1}' + y_n' \right) \qquad (4.35)$$

Subtracting Equation (4.33) from Equation (4.35) and simplifying, we get,

$$y_{n+1} - y(t_{n+1}) = \frac{\Delta t_n^3}{12} y_n''' + O(\Delta t_n)^4 \qquad (4.36)$$

Subtracting Equation (4.34) from Equation (4.36) and eliminating the unknowns *i.e.*, y_n''' and $y(t_{n+1})$, we obtain,

$$y_{n+1} - y_{n+1}^p = 3 \left(\frac{\Delta t_n^3}{12} y_n''' \right) \left(1 + \frac{\Delta t_{n-1}}{\Delta t_n} \right) + O(\Delta t_n)^4 \qquad (4.37)$$

we now define,

$$y_{n+1} - y(t_{n+1}) = d(y_{n+1})$$

and substitute its value from Equation (4.36) in Equation (4.37), to obtain,

$$d(y_{n+1}) = \frac{y_{n+1} - y_{n+1}^p}{3\left(1 + \frac{\Delta t_{n-1}}{\Delta t_n}\right)} + O(\Delta t_n)^4 \qquad (4.38)$$

From Equation (4.36), we can write for times t_{n+1} and t_{n+2},

$$\frac{d(y_{n+2})}{d(y_{n+1})} = \left(\frac{\Delta t_{n+1}}{\Delta t_n}\right)^3 \left(\frac{y_{n+1}'''}{y_n'''}\right)$$

but,

$$y_{n+1}''' = y_n''' + O(\Delta t_n)$$

therefore,

$$\frac{d(y_{n+2})}{d(y_{n+1})} = \left(\frac{\Delta t_{n+1}}{\Delta t_n}\right)^3 + \left(\frac{\Delta t_{n+1}}{\Delta t_n}\right)^3 \left(\frac{O(\Delta t_n)}{y_n'''}\right) \qquad (4.39)$$

If we specify an acceptable error ϵ for $|\, d(y_{n+2})\, |$ and neglect the higher order term in Equation (4.39), we can solve for the next time-step Δt_{n+1}, as,

$$\Delta t_{n+1} = \Delta t_n \left(\frac{\epsilon}{|\, d(y_{n+1})\, |}\right)^{\frac{1}{3}} \qquad (4.40)$$

The norm used is as suggested by Gresho et. al. [9]

$$|\, d(y_{n+1})\, | = \sqrt{\frac{1}{N}\left[\frac{1}{T_{max}^2}\sum_{i=1}^{N} d_i^2(T_{n+1})\right]} \qquad (4.41)$$

Where N is the total number of nodal variables and $\mathbf{T}$ are the nodal variables.

Bixler [7] suggests some improvements in the scheme as described above. He suggests the use of the mid point rule instead of the trapezoidal rule as corrector while keeping the AB formula as predictor. The midpoint rule for the equation $y' = f(y, t)$ can be written as,

$$\begin{aligned}
y_{n+1} &= y_n + \Delta t_n \left(f_{n+\frac{1}{2}}\right) \\
&= y_n + \Delta t_n \left(y_{n+\frac{1}{2}}'\right) \qquad (4.42)
\end{aligned}$$

Bixler calculates the time truncation error for the midpoint rule by subtracting the Taylor series expansion from Equation (4.42), as,

$$y_{n+1} - y(t_{n+1}) = -\frac{\Delta t_n^3}{24}\left(y_n''' - \frac{3\,(y_n'')^2}{y_n'}\right) + O(\Delta t_n)^4 \qquad (4.43)$$

In calculating the time derivative y' for use in the predictor, he suggests a formula based on the values of y at different time levels, to improve the stability of the predictor, *i.e.*,

$$y'_n = \frac{2}{\Delta t_{n-1}}(y_n - y_{n-1}) - y'_{n-1} \qquad (4.44)$$

where,

$$y'_{n-1} = \frac{\Delta t_{n-2}}{\Delta t_{n-1} + \Delta t_{n-2}}\left(\frac{y_n - y_{n-1}}{\Delta t_{n-1}}\right) + \frac{\Delta t_{n-1}}{\Delta t_{n-1} + \Delta t_{n-2}}\left(\frac{y_{n-1} - y_{n-2}}{\Delta t_{n-2}}\right)$$
$$(4.45)$$

The formula for the succeeding time-step remains the same except, $d(y_{n+1})$ is redefined for the new corrector, *i.e.*,

$$d(y_{n+1}) = \frac{\gamma}{2 + \gamma + 3\frac{\Delta t_{n-1}}{\Delta t_n}}(y_{n+1} - y^p_{n+1}) \qquad (4.46)$$

where, $0.25 \leq \gamma \leq 1.0$. with $\gamma = 1$, the original formula is obtained. This scheme helps to increase the time-step much more rapidly when steady state impends.

For both the above time integration methods the AB formula is used as predictor which requires the time derivative, $\dot{\mathbf{T}}_o$ at the first step, which can be obtained by solving Equation (4.5) as,

$$\mathbf{C}_o\dot{\mathbf{T}}_o = \mathbf{F}_o - \mathbf{K}_o\mathbf{T}_o \qquad (4.47)$$

The same procedure may be used to obtain time-step selection for the simpler first order methods. If the Forward Euler method is used as a predictor and the Backward Euler as the corrector, the formula for the succeeding time-step is written as,

$$\Delta t_{n+1} = \Delta t_n\left(\frac{\epsilon}{\mid d(y_{n+1})\mid}\right)^{\frac{1}{2}} \qquad (4.48)$$

where,

$$d(y_{n+1}) = \frac{y_{n+1} - y^p_{n+1}}{2} + O(\Delta t_n)^3 \qquad (4.49)$$

This is only for linear case.

4.5   Benchmark Example

An example to test the transient algorithms as described in the previous sections was obtained from a list of selected benchmarks published by NAFEMS([10]). Figure 4.1 shows the details of the problem.

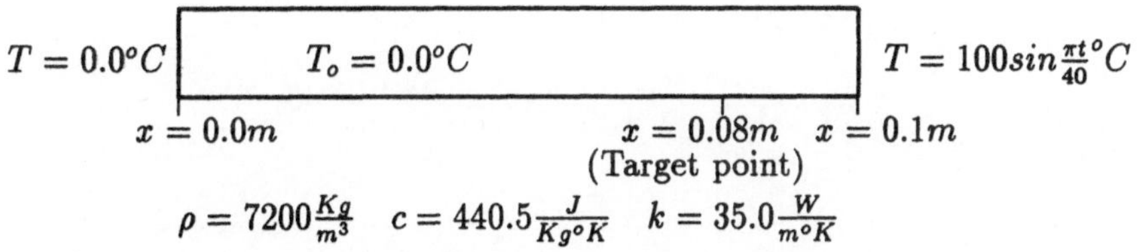

$$T = 0.0°C \boxed{\qquad T_o = 0.0°C \qquad\qquad} T = 100sin\frac{\pi t}{40}°C$$

$$x = 0.0m \qquad\qquad x = 0.08m \quad x = 0.1m$$

(Target point)

$$\rho = 7200\frac{Kg}{m^3} \quad c = 440.5\frac{J}{Kg°K} \quad k = 35.0\frac{W}{m°K}$$

Figure 4.1: Benchmark example for transient analysis

As can be seen from the figure, this is a one dimensional heat conduction problem with one end at a fixed temperature of $0.0°C$ and the other end subjected to an oscillating boundary condition. The material properties correspond to those of steel, with ρ, c and k being the density, specific heat and thermal conductivity. The temperature at time $t = 0.0$ (initial condition) is zero.

The analytical solution to this problem is available in reference [11] and is reproduced in the following lines. The temperature T at time t for problems of the type shown in Figure 4.1 is given by,

$$T = \frac{2}{L}\sum_{n=1}^{\infty}\frac{nk'\pi A}{L(\omega^2 + \alpha_n^2)}sin\frac{n\pi x}{L}(\alpha_n sin\omega t - \omega(cos\omega t - e^{-\alpha_n t})) \qquad (4.50)$$

for $0 < x < L$, L being the length. The boundary conditions are given by,

$$T = Asin\omega t \qquad \text{at } x = 0$$

and

$$T = 0 \qquad \text{at } x = L$$

Also, values of α_n and k' are obtained from,

$$\alpha_n = \frac{k'n^2\pi^2}{L^2}$$

and

$$k' = \frac{k}{\rho c}$$

Figure 4.2 shows the plot of the analytical solution for the chosen benchmark problem at $x = 0.08m$. The exponential term of Equation (4.50) (shown by the dotted line in the Figure) corresponds to the transient component which decays to a negligible value for larger values of t and a steady harmonic component remains. The finite element method using a uniform mesh of 20 6-noded elements produces a result indistinguishable from the analytical solution. The timestep size was $2secs$ with $\alpha = 0.5$.

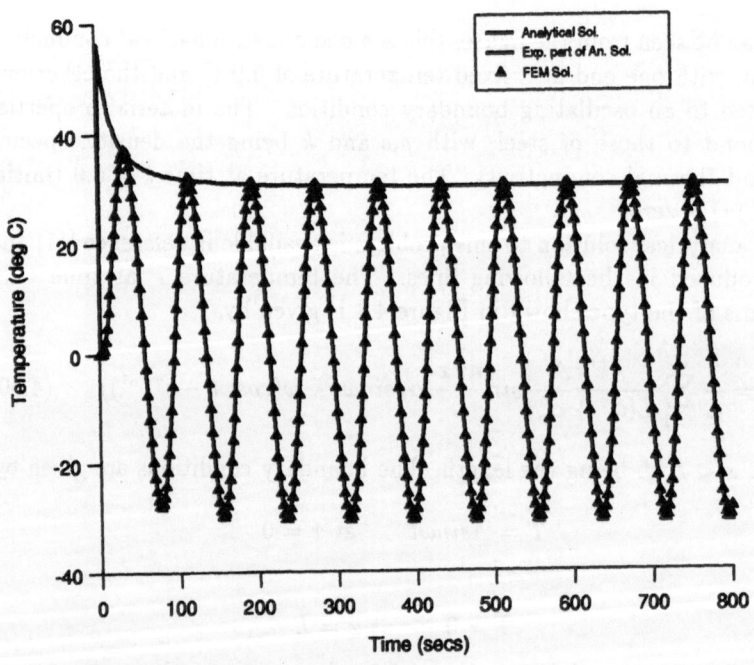

Figure 4.2: Analytical and FEM solution for the benchmark problem

References

[1] J.R.Yu and T.R.Hsu. Analysis of heat conduction in solid by space-time finite element method. *International Journal for Numerical Methods in Engineering*, 21:2001–2012, 1985.

[2] R.W.Lewis J.Y.Zhang and J.S.Peng. A semi-analytic approach to general transient problems and its applications to heat transfer. *Numerical Heat Transfer*, 23:413–424, 1993.

[3] T.J.R.Hughes. *The Finite Element Method - Linear Static and Dynamic Finite Element Analysis*. Prentice-Hall International, Inc., Englewood Cliffs, New Jersey 07632, 1987.

[4] T.J.R.Hughes. Analysis of transient algorithms with particular reference to stability behaviour. In *Computaional methods for transient analysis*. Elsevier Science Publishers, 1983.

[5] W.L.Wood and R.W.Lewis. A comparison of time marching schemes for the transient heat conduction equation. *International Journal for Numerical Methods in Engineering*, 9:679–689, 1975.

[6] G.Dahlquist and B.Lindberg. On some implicit one-step methods for stiff differential equations. Technical Report TRITA-NA-7302, Department of Information Processing, The Royal Institute of Technology, Stockholm, 1973.

[7] N.E.Bixler. An improved time integrator for finite element analysis. *Commnunications in Applied Numerical Methods*, 5:69–78, 1989.

[8] K.A.Cliffe. On conservative finite element formulations of the inviscid boussinesq equations. *International Journal for Numerical Methods in Fluids*, 1:117–127, 1981.

[9] P.M.Gresho, R.L.Lee, and R.L.Sani. On the time-dependent solution of the incompressible Navier-Stokes equations in two and three dimensions. In *Recent Advances in Numerical Methods in Fluids*, volume 1. Pineridge Press Limited, Swansea, 1980.

[10] J.Barlow and G.A.O.Davies. Selected FE benchmarks in structural and thermal analysis. Technical Report FEBSTA REV 1, NAFEMS, 1986.

[11] H.S.Carslaw and J.C.Jaeger. *Conduction of Heat in Solids*. Clarendon Press, Oxford, 1959.

Chapter 5

Phase Transformation

5.1   Introduction

Phase transformations take place in familiar ways such as solidification, melting, vaporisation and condensation etc. A direct transformation from solid to vapour phase is called sublimation. From a thermodynamics point of view, when a system consists of more than one phase, each phase may be considered as a seperate system within the whole. The thermodynamic parameters of the whole system may then be constructed from those of the component phases. However, assuming thermodynamic equilibrium and restricting the interaction between the phases at the interface, solely to the flow of heat, no new complications arise and the two phase system can essentially be treated as a single system with anisotropic properties. This assumption has been used in this chapter to develop the algorithms for the numerical simulation of phase transformation processes. Furthermore, the processes considered here have been restricted to melting and solidification, involving only the solid and liquid phases of materials. The assumptions made above cease to remain valid, when the vapour phase is considered.

The most familiar phase change process is undoubtedly the melting of ice and the solidification of water. The modelling of phase change may be required for many not so familiar engineering problems, such as, the analysis of nuclear fuel containers, and other areas of nuclear power generation such as accident analysis. Some solar energy storage devices use phase change materials to store thermal energy in the form of latent heat. Geotechnical applications include soil freezing for excavations etc., construction over permafrost [1, 2], rock formation by freezing of volcanic eruptions, igneous intrusions (in formation of igneous rocks). Glass forming involves phase transformation but negligible latent heat is involved. In the food industry the freezing of meat, freeze drying of coffee etc. are areas of phase change modelling applications. In medicine the technique of cryosurgery (freezing and removal of malignant tissue) involves phase change modelling. Semiconductor crystal growth is another application.

The analysis of fire resisting containers which incorporate phase change materials with high latent heats to absorb heat in melting and thus delay any damage to contents, is another example [3]. However, the single most important application of phase change modellling involves materials processing, especially metals. These include purification of metals, growth of pure crystals from melts and solutions, solidification of castings and ingots, welding, electro slag melting, zone melting etc.

Melting and solidification are phase transformation processes which are accompanied by either absorption or release of thermal energy. A moving boundary exists that separates the two phases of differing thermo-physical properties and at which thermal energy is absorbed or liberated. As an example, if we consider the solidification of a casting or ingot, the super-heat in the melt and the latent heat of fusion liberated at the solid-liquid interface are transferred across the solidified metal, the metal/mould interface and the mould encountering at each of these stages a certain thermal resistance. During solidification of binary and multicomponent alloys the physical phenomena become more complicated due to the phase transformation taking place over a range of temperatures. The lowest temperature corresponding to the fully liquid phase is is called the 'liquidus' and the highest temperature corresponding to the solid phase is called the 'solidus'. These temperatures vary according to the concentrations of the various alloy components, which can be seen from the phase diagram of a particular alloy. During the solidification of an alloy the concentrations vary locally from the original mixture as material may be preferentially incorporated or rejected at the solidification front. The material between the solidus and liquidus temperatures is partly solid and partly liquid and resembles a porous medium and is referred to as a *mushy* zone.

As is evident from the above discussion, a complete understanding of the phase change phenomenon involves the analysis of the various processes that accompany it. The most important of these processes from a macroscopic point of view, with which this work is concerned, is the heat transfer process. This is complicated by the release or absorption of the latent heat of fusion at the liquid/solid interface. Several methods have been used to take account of the latent heat within different thermal analysis programs. This chapter is devoted to a review and discussion of these methods. These methods are generally divided into fixed and moving mesh methods and are also referred to as the 1-domain and the 2-domain methods, respectively. As the names suggest 1-domain methods involve the solution of a continuous system with an implicit representation of the phase change while in the 2-domain or 'front tracking' methods the solid and the liquid regions are treated seperately and the phase change interface is explicitly determined as a moving boundary. A comprehensive treatment of the subject of moving boundary problems in general and phase change related moving boundary problems in particular appears in a book by John Crank [4]. A good account of the early effort in the numerical modelling of phase change (primarily solidification) is given by Samonds [5], who reports that the finite difference method was first used by Sarjant and Slack [6] in 1954 in the study of ingot solidification. Many other researchers continued to employ the finite difference methods to predict solidifcation behaviour using computers, such as Henzel and Kevarian [7, 8, 9], Weatherwax and Riegger [10], Hansen [11] etc. The first use of the finite element method was by Soliman and Fakhroo [12] in the analysis of ingot casting, as reported in reference [5]. Morgan *et. al.* and Lewis *et. al.* [13, 14] modelled the solidification of ingot castings and also analysed the thermal stresses in the ingot. A useful review of the methods used in a finite element context, can be found in a paper by Dalhuijsen and Segal [15]. Salcudean and Abdullah [16] have listed several latent heat formulations and many references for general numerical modelling of phase change. Recently Voller *et al.* [17] have discussed the fixed grid methods with the aim of providing a comprehensive review of the major formulations with an emphasis on the numerical features.

5.2 The Stefan Problem

The solidification problem is sometimes referred to as the *Stefan problem* due to the fact that the first published discussion of this problem was by Stefan, who studied the thickness of the polar ice cap. Stefan's solution was a special case of the more general Neumann's solution as it assumed the liquid to be at its melting point. The classical problem involves considering the conservation of energy in the domain Ω by dividing it into two distinct domains Ω_l and Ω_s, where, $\Omega_l + \Omega_s = \Omega$. The energy conservation is written as,

$$\rho_l c_l \frac{\partial T}{\partial t} = \nabla \cdot k_l \nabla T \qquad \text{in } \Omega_l \qquad (5.1)$$

and

$$\rho_s c_s \frac{\partial T}{\partial t} = \nabla \cdot k_s \nabla T \qquad \text{in } \Omega_s \qquad (5.2)$$

where, the subscripts l and s denote liquid and solid respectively. The complete description of the problem involves, in addition to the initial conditions and the appropriate external boundary conditions, the interface conditions on the phase change boundary Γ_{sl}, which are,

$$T_{\Gamma_{sl}} = T_f$$

$$k_l \left(\frac{\partial T}{\partial x}\right)_l - k_s \left(\frac{\partial T}{\partial x}\right)_s = \rho L \frac{ds}{dt} \qquad \text{on } \Gamma_{sl} \qquad (5.3)$$

where, s represents the position of the interface, $\frac{ds}{dt}$ the interface velocity and T_f is the phase change temperature. The main complication in solving the classical problem lies in tracking the interface boundary position s.

5.3   Numerical Methods for Modelling Phase Transformation

As mentioned earlier the front tracking methods attempt to solve the classical problem in its pure or strong form. This involves the use of moving meshes or the transformation of coordinates [18, 19, 20] to explicitly satisfy the interface conditions. These methods can be used to solve isothermal phase change problems with good accuracy but become too complicated, even impossible, when faced with complex interface shapes which vary nonmonotonically with time. Furthermore, front tracking methods cannot be readily used in the case of freezing over a range of temperatures. Due to the limitations of the 2-domain methods the subsequent discussion will be limited to the 1-domain methods. The 1-domain or fixed grid methods offer a more general solution as they account for the phase change conditions implicitly without attempting to *a priori* establish the position of the front. These methods are based on a weak formulation of the classical problem, which is commonly referred to as the *enthalpy formulation*. A single, energy conservation equation is written for the whole domain as,

$$\frac{\partial H}{\partial t} = \nabla \cdot k \nabla T \quad \text{in } \Omega \tag{5.4}$$

where, H is the enthalpy function or the total heat content which is defined in reference [15] for isothermal phase change as,

$$H(T) = \int_{T_r}^{T} \rho c_s(T) dT \qquad (T < T_f)$$

$$H(T) = \int_{T_r}^{T_f} \rho c_s(T) dT + \rho L + \int_{T_f}^{T} \rho c_l(T) dT \qquad (T \geq T_f) \tag{5.5}$$

and for phase change over an interval of temperatures T_s to T_l, which are the solidus and the liquidus respectively, we have,

$$H(T) = \int_{T_r}^{T} \rho c_s(T) dT \qquad (T < T_s)$$

$$H(T) = \int_{T_r}^{T_s} \rho c_s(T) dT + \int_{T_s}^{T} \left(\rho \left(\frac{dL}{dT} \right) + \rho c_f(T) \right) dT \qquad (T_s \leq T \leq T_l)$$

$$H(T) = \int_{T_r}^{T_s} \rho c_s(T) dT + \rho L + \int_{T_s}^{T_l} \rho c_f(T) dT + \int_{T_l}^{T} \rho c_l(T) dT \qquad (T > T_l)$$

$$\tag{5.6}$$

where, c_f is the specific heat in the freezing interval, L is the latent heat and T_r is a reference temperature lower than T_s. The mathematical validity of this formulation is discussed in detail by Crank [4], who has referred to several original works, such as that of Rubinstein, Kamenomostskaja,

Oleinik, Niezgodka etc. to state the equivalence of the classical and weak formulations and the existence and uniqueness of the solution for the weak formulation.

The enthalpy formulation can be implemented in a finite element based heat transfer code in a variety of ways. Some of the commonly used methods are discussed in the following text.

5.3.1 Effective Heat Capacity

Among the fixed mesh methods, one of the earliest and the most commonly used methods has been the *effective heat capacity* method. This method was derived from writing Equation (5.4) as,

$$\frac{dH}{dT}\frac{\partial T}{\partial t} = \nabla \cdot k \nabla T \qquad \text{in } \Omega \tag{5.7}$$

Comparing with the standard heat conduction equations such as Equations (5.1) and (5.2), we can write,

$$c_{eff} = \frac{dH}{dT} \tag{5.8}$$

where, c_{eff} is the effective heat capacity, which can be evaluated directly from Equations (5.6) as shown in reference [15],

$$
\begin{aligned}
c_{eff} &= \rho c_s & (T < T_s) \\
c_{eff} &= \rho c_f + \frac{L}{T_l - T_s} & (T_s \leq T \leq T_l) \\
c_{eff} &= \rho c_l & (T > T_l)
\end{aligned}
\tag{5.9}
$$

Figure 5.1 shows the typical variation of the effective heat capacity and enthalpy with temperature. It can be seen from the figure that if this directly evaluated effective specific heat is used, it will be necessary to maintain an interval of temperatures for the evolution of latent heat, otherwise the effective heat capacity will become infinite. This method therefore cannot accurately model an iso-thermal change of phase due to the requirement of a temperature range. In reality, the use of a directly evaluated heat capacity is of limited value in cases other than those where the phase change occurs in a very wide range of temperatures. The main problem in the numerical modelling context is that the latent heat information may be contained in a narrow band of temperatures, and it may be missed if the temperature change in one time step at a node or integration point straddles the phase change interval [5, 15, 21]. This imposes severe restrictions on the temporal and spatial step sizes in a numerical analysis. Due to the step like behaviour of c_{eff} around the phase change interval, numerical oscillations may occur, making the achievement of a convergent solution

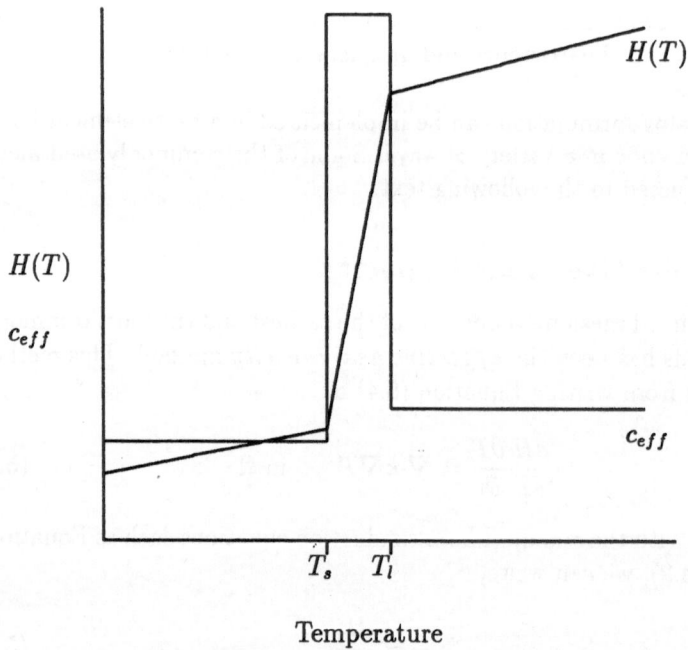

Figure 5.1: Typical variation of enthalpy and c_{eff} with temperature

difficult [15]. Lewis and Huang [22] have used an effective capacity type
method described by Samonds in [5] to analyse a fire resisting cabinet containing phase change material. As the amount of latent heat was large,
they had to use an adaptive meshing procedure, with an extremely fine
mesh in the region of the moving phase change front.

5.3.2   Enthalpy Method

In order to overcome the difficulties encountered in using a directly evaluated effective capacity, recourse is made to several averaging techniques.
These techniques are generally referred to as the enthalpy method. In this
context the enthalpy method enables the heat capacity to be defined as a
smooth function of temperature [21]. Some of the commonly used averaging techniques mentioned in the relevent references [15, 21, 23] are listed
below. The simplest approximation used (for 2-D) is,

$$\frac{dH}{dT} = \frac{1}{2}\left(\frac{\frac{\partial H}{\partial x}}{\frac{\partial T}{\partial x}} + \frac{\frac{\partial H}{\partial y}}{\frac{\partial T}{\partial y}}\right) \qquad (5.10)$$

This method has been reported to cause oscillations in certain circumstances [21]. A better method as reported by DelGuidice *et al.* [1] is,

$$\frac{dH}{dT} = \left(\frac{\left(\frac{\partial H}{\partial x}\right)\left(\frac{\partial T}{\partial x}\right) + \left(\frac{\partial H}{\partial y}\right)\left(\frac{\partial T}{\partial y}\right)}{\left(\frac{\partial T}{\partial x}\right)^2 + \left(\frac{\partial T}{\partial y}\right)^2} \right) \tag{5.11}$$

Morgan *et al.* [21] and Lemmon [24] suggest another approximation which is reported to work satisfactorily,

$$\frac{dH}{dT} = \left(\frac{\left(\frac{\partial H}{\partial x}\right)^2 + \left(\frac{\partial H}{\partial y}\right)^2}{\left(\frac{\partial T}{\partial x}\right)^2 + \left(\frac{\partial T}{\partial y}\right)^2} \right)^{\frac{1}{2}} \tag{5.12}$$

Morgan *et al.* [21] have also advocated the use of a simple backward difference approximation,

$$\left(\frac{dH}{dT} \right)_n = \frac{(H_n - H_{n-1})}{(T_n - T_{n-1})} \tag{5.13}$$

where, n represents the timestep number. This scheme restricts the time step severely according to reference [15] if a correct heat balance is to be maintained. Lewis and Roberts [23] claim that the last scheme is computationally quicker than the abovementioned averaging techniques.

In using the above techniques in a finite element analysis it is common practice to interpolate H from the nodal values using the same basis functions as for T thus obtaining a smoothing effect. The appearance of space derivatives in the above equations ensures the inclusion of the phase change effect. Dalhuijsen and Segal [15] have compared the performance of the various approximation schemes mentioned above with respect to time integration procedures, and found that the method of DelGuidice *et al.* is marginally more suitable for Euler backward time integration and Lemmon's method works slightly better for two-step time integration schemes, but in general there is no significant effect on the relative accuracy. They also advocate the use of lumped mass matrices for better accuracy and reduction in the tendency for oscillation.

The enthalpy method, although reasonably accurate and very simple to implement in any heat transfer code, suffers from a number of deficiencies some of which have been discussed in [25, 26]. The main drawbacks in a finite element context can be listed as follows,

- Due to the necessity of a temperature range for the evolution of latent heat, isothermal phase changes can not be modelled correctly.

- Requires small spatial and temporal step sizes, otherwise convergence becomes difficult. This does not allow for quick computer runs if a rough preliminary analysis is required.

- Due to the requirement of small mesh size and timesteps an analysis can become prohibitively expensive if a difficult problem is encountered, such as that with a narrow freezing range and large latent heat.

- The actual amount of latent heat absorbed/released during a phase change analysis can at best be approximate, as it is accounted for in an indirect manner.

5.3.3   Heat Source Method

A fixed mesh method which is gaining favour (and is essentially an enthalpy based method as well), is the *Fictitious Heat − Flow Method* first suggested in the finite element context by Rolph and Bathe [27] and then by Roose and Storrer [28]. This is also referred to as the *heat source method*, as will be done here. This method lacks the mathematical theory which supports the enthalpy method so well, and also the simplicity with which the enthalpy method can be implemented in finite element heat transfer programs. In spite of these drawbacks it offers an attractive alternative to the enthalpy method, as it does not suffer from any of the deficiencies of the enthalpy method listed above. The central idea behind the source method is to lump all the latent heat available at the nodes, which may then be released/absorbed as an internal heat source at the appropriate temperature or a range of temperatures. This enables the modelling of isothermal phase changes as well as phase changes over a temperature range. From examining Equations (5.4),(5.5) and (5.6) we can see that the rate of change of enthalpy in Equation (5.4) can be broken into sensible and latent heat parts. Subsequently, the latent heat part can be used as the internal heat source or sink term Q in the energy Equations (2.22) and (2.30) in Chapter 2. Therefore we can write,

$$Q = \rho \frac{d}{dt} \left(\int_\Omega L d\Omega \right) \tag{5.14}$$

The implementation of this method in a finite element heat transfer program is not straightforward. Rolph and Bathe [27] have given considerable detail for such an implementation. In the following lines a step by step procedure of implementing source based phase change in a finite element program is presented.

1. Calculate the global nodal latent heat vector $QTOTL(i)$ for each node i in the finite element mesh. This can be done by lumping the masses of all elements at the nodes using standard lumping procedures [29] and assembling all nodal contributions in the vector $QTOTL(i)$ after multiplying them by the latent heat per unit volume. Also store the assembled nodal masses in $QMASS(i)$.

2. Set up a vector $QCUMU(i)$ for collecting the amount of latent heat released or absorbed for each node i in the course of the analysis. The value of $QCUMU(i)$ is initialised to zero before the start of the analysis. Another vector $QLATH(i)$ is set up for use in the global load vector as the fictitious heat-flow or the heat source vector. The value of this vector is initialised to zero at the beginning of each timestep. Lastly, a vector $QINCR(i)$ is set up for storing the increments in the latent heat in each iteration. The value of this vector is initialised to zero at the beginning of each iteration in a single time step.

3. At the start of the analysis flag all nodes as solid or liquid using $ISOLID(i)$. In the program developed for this work the flag value -1 indicates liquid, $+1$ indicates solid and 0 indicates a node in the process of changing phase. Another flag $LSOLID(i)$ is used to indicate melting or solidification. A negative value of the flag indicates solidification and a positive value indicates melting.

4. As the problem is nonlinear and the latent heat is released or absorbed according to the rise and fall in temperature in each iteration, at least two or more iterations will be required for each timestep. For a typical timestep n and iteration p it is required to check the solid and liquid nodes or nodes for which $ISOLID(i) \neq 0$ that,

$$\left(T_n^{p-1}\right)_i > T_l > (T_n^p)_i \qquad \text{(solidification)}$$

or,

$$\left(T_n^{p-1}\right)_i < T_s < (T_n^p)_i \qquad \text{(melting)}$$

If either of the above conditions are true then $ISOLID(i)$ is set to 0. The latent heat to be released or absorbed in this case is calculated as,

$$QINCR(i) = \hat{c}\,(T_l - (T_n^p)_i)\,QMASS(i) \qquad \text{(for solidification)}$$

or,

$$QINCR(i) = \hat{c}\,(T_s - (T_n^p)_i)\,QMASS(i) \qquad \text{(for melting)}$$

It may be noted that in the first case $QINCR(i)$ will be positive while in the second case it will be negative. If the node under consideration is already undergoing phase change from the previous timestep $n-1$, so that $ISOLID(i) = 0$, then the latent heat is calculated for both solidification and melting as,

$$QINCR(i) = \hat{c}\,((T_{n-1})_i - (T_n^p)_i)\,QMASS(i)$$

5. The incremental latent heat is added to the heat source vector $QLATH(i)$ for the time step n, which is then added to the force vector as in Equation (2.32) in chapter 2, after dividing by the timestep Δt_n as the rate of latent heat release/absorption is required. It may be noted that this technique allows for the use of variable timesteps. The iterations are continued until convergence.

6. After each iteration, the temperature of each node undergoing phase change is corrected (assuming a latent heat evolution linear with temperature) to keep it consistent with the amount of latent heat released or absorbed according to,

$$(T_n^p)_i = T_l - \frac{|QCUMU(i)|}{QTOTL(i)}(T_l - T_s) \qquad \text{(for solidification)}$$

or,

$$(T_n^p)_i = T_s + \frac{|QCUMU(i)|}{QTOTL(i)}(T_l - T_s) \qquad \text{(for melting)}$$

7. The incremental latent heat $QINCR(i)$ is also added to the cumulative latent heat vector $QCUMU(i)$ for each iteration of all time steps until,

$$QCUMU(i) \geq QTOTL(i)$$

when the node i is flagged as solid or liquid as the case may be.

The technique described above is in terms of the more general case of mushy phase change involving a solidus and a liquidus temperature (T_s and T_l), the special case of isothermal phase change can be derived by making $T_s = T_l$ in all the above equations. The $\hat{c}$ term used in the fourth step above, is an effective specific heat [27]. The value of $\hat{c}$ is the same as c_s or c_l mentioned in Equation (5.5) for isothermal phase change. For mushy phase change the value of $\hat{c}$ is calculated by considering the fact that the heat lost/gained in one iteration is the sum of the latent heat and the sensible heat and the enthalpy-temperature curve is linear between liquidus and solidus. If we consider a drop in temperature of ΔT in one iteration then the total heat loss is,

$$c_f \Delta T$$

where, c_f is the same as in Equation (5.6). This includes the heat loss due to latent heat release and due to the corrected temperature drop δT. Assuming linear latent heat release we can write for the heat loss due to latent heat,

$$c_f(\Delta T - \delta T) = L\left(\frac{\delta T}{T_l - T_s}\right) \qquad (5.15)$$

If an effective specific heat ($\hat{c}$) is defined so that it directly gives the heat loss due to latent heat (the r.h.s of the above equation), then we can write,

$$\hat{c}\Delta T = \left(\frac{L}{T_l - T_s}\right)\delta T \tag{5.16}$$

solving for δT from Equation (5.15), we have

$$\delta T = \frac{c_f\Delta T}{\left(c_f + \frac{L}{T_l - T_s}\right)} \tag{5.17}$$

substituting Equation (5.17) into Equation (5.16) we obtain the final expression for $\hat{c}$ as,

$$\hat{c} = \frac{1}{\left(\frac{T_l - T_s}{L} + \frac{1}{c_f}\right)} \tag{5.18}$$

The same relation can be obtained if a temperature rise (melting) is considered instead of a temperature drop.

It can be noted from the above description that although this method is more complicated to implement, it is more attractive in terms of engineering intuition, as the exact amount of latent heat can be explicitly released/absorbed at the appropriate temperature or range of temperatures. Comparisons with other methods will be made by solving a benchmark example in the succeeding section.

5.4   Benchmark Examples

A one dimensional benchmark example of solidification as shown in Figure 5.2 is solved by the methods discussed above and compared with the analytical solution. The material properties, dimensions, boundary and initial conditions are as shown in the same figure. T_o and T_f refer to the initial and freezing temperatures respectively. No particular units are necessary. This example has been used for such comparisons in several papers [21, 27, 28]. The temperature vs. time curve at a chosen point ($x = 1.0$ here) in the domain is used for the comparisons. This method is more reliable according to [15], instead of the normally used interface position vs. time curves. The same example with reversed temperatures as shown in Figure 5.3 will be used to solve a melting example.

The mesh with 20 9-noded elements used to solve the above examples is as shown in Figure 5.4. This mesh is used for all cases except for one case where the source method is demonstrated on a coarse mesh.

5.4.1   Analytical Solution

The general solution to Neumann's problem and many special cases appear in the textbook by Carslaw and Jaeger [30] and that by Crank [4].

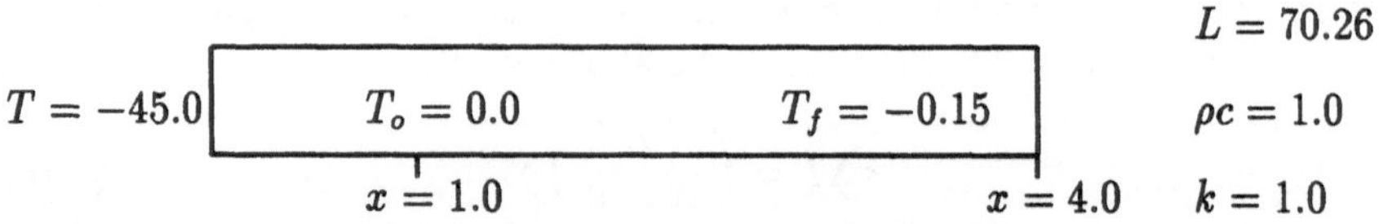

$$L = 70.26$$

| $T = -45.0$ | $T_o = 0.0$ | $T_f = -0.15$ | $\rho c = 1.0$ |

$$x = 1.0 \qquad\qquad x = 4.0 \qquad k = 1.0$$

Figure 5.2: Solidification example.

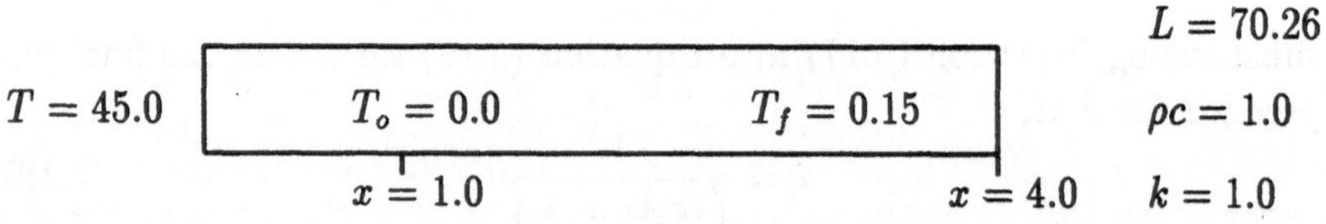

$$L = 70.26$$

| $T = 45.0$ | $T_o = 0.0$ | $T_f = 0.15$ | $\rho c = 1.0$ |

$$x = 1.0 \qquad\qquad x = 4.0 \qquad k = 1.0$$

Figure 5.3: Melting example.

The analytical solutions for 1-D solidification and melting were taken from [30] and were implemented on the computer using a bisection method to generate temperature time curves at specified points in the domain.

Analytical Solution for Solidification

The anlytical solution for a semi-infinite 1-D domain at a temperature T_∞, subjected to a surface temperature of zero degrees, is calculated according to the following formulae:

The position of the solidification front X is obtained from,

$$X = 2\lambda(k_s t)^{\frac{1}{2}} \tag{5.19}$$

The temperature in the solid zone ($x \leq X$ or $T \leq T_f$), is given by,

$$T = \frac{T_f}{erf\lambda}erf\frac{x}{2(k_s t)^{\frac{1}{2}}} \tag{5.20}$$

The temperature in the liquid zone ($x \geq X$ or $T \geq T_f$), is given by,

$$T = T_\infty - \frac{T_\infty - T_f}{erfc\lambda(\frac{k_s}{k_l})^{\frac{1}{2}}}erfc\frac{x}{2(k_l t)^{\frac{1}{2}}} \tag{5.21}$$

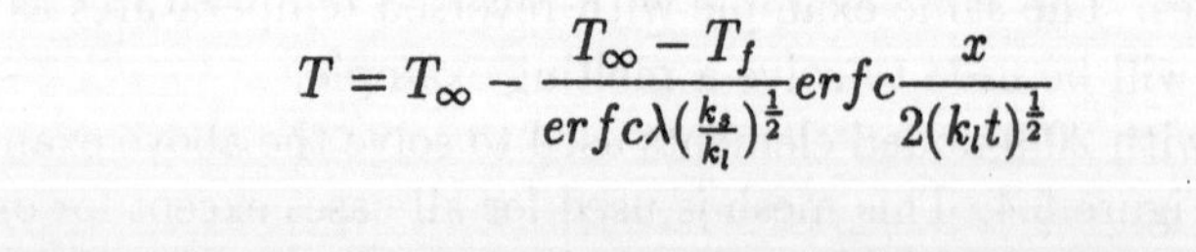

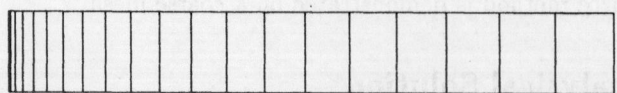

Figure 5.4: Finite element mesh for the benchmark examples

λ in the above equations is obtained from,

$$\frac{e^{-\lambda^2}}{erf\lambda} - \left(\frac{k_l}{k_s}\right)^{\frac{1}{2}} \frac{T_\infty - T_f}{T_f} \frac{e^{-\lambda^2 \frac{k_s}{k_l}}}{erfc\lambda(\frac{k_s}{k_l})^{\frac{1}{2}}} = \frac{\lambda L \pi^{\frac{1}{2}}}{c_s T_f} \tag{5.22}$$

Here, erf is the error function and $erfc(x) = 1 - erf(x)$. The values of the error function were approximated from the formula below [31]:

$$erf(x) = 1 - (a_1 b + a_2 b^2 + a_3 b^3 + a_4 b^4 + a_5 b^5)e^{-x^2} + \epsilon(x) \tag{5.23}$$

where,

$$b = \frac{1}{1 + px}$$

$$p = 0.3275911$$

$$a_1 = 0.254829592$$

$$a_2 = -0.284496736$$

$$a_3 = 1.421413741$$

$$a_4 = -1.453152027$$

$$a_5 = 1.061405429$$

$$\epsilon(x) \leq 1.5 \times 10^{-7}$$

Analytical Solution for Melting

The anlytical solution for a semi-infinite 1-D domain at a temperature of zero degrees, subjected to a surface temperature of T_o, is calculated according to the following formulae:

The position of the melting front X is obtained from,

$$X = 2\lambda(k_l t)^{\frac{1}{2}} \tag{5.24}$$

The temperature in the liquid zone ($x \leq X$ or $T \geq T_f$), is given by,

$$T = T_o - \frac{T_o - T_f}{erf\lambda} erf \frac{x}{2(k_l t)^{\frac{1}{2}}} \tag{5.25}$$

The temperature in the solid zone ($x \geq X$ or $T \leq T_f$), is given by,

$$T = \frac{T_f}{erfc\lambda(\frac{k_l}{k_s})^{\frac{1}{2}}} erfc \frac{x}{2(k_s t)^{\frac{1}{2}}} \tag{5.26}$$

λ in these equations is obtained from,

$$\frac{e^{-\lambda^2}}{erf\lambda} - \left(\frac{k_s}{k_l}\right)^{\frac{1}{2}} \frac{T_f}{T_o - T_f} \frac{e^{-\lambda^2 \frac{k_l}{k_s}}}{erfc\lambda(\frac{k_l}{k_s})^{\frac{1}{2}}} = \frac{\lambda L \pi^{\frac{1}{2}}}{c_l(T_o - T_f)} \tag{5.27}$$

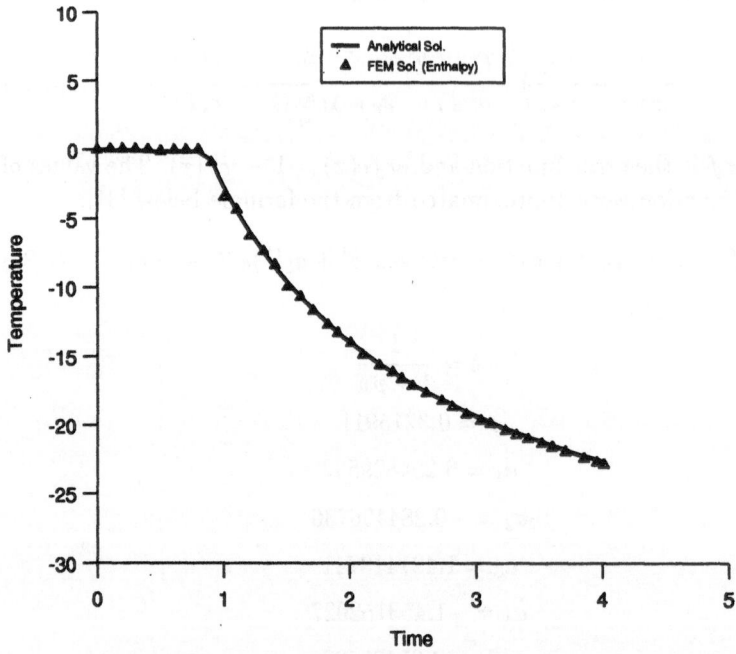

Figure 5.5: Enthalpy method vs the analytical solution.

5.4.2 Numerical Solution

In addition to the heat source method, the enthalpy method of Morgan *et al.* [21] as given by Equation (5.13) has been used. An effective capacity method used by Samonds [5] has also been compared.

Figures 5.5 and 5.6 show the comparisons of the enthalpy and the source methods with the analytical solution for the 1-D solidification problem. Although the enthalpy method appears to be relatively more accurate, roughly 1700 timesteps were required to obtain this solution, while only 60 timesteps were used by the source method to reach the same time level, this number could have been even smaller if the maximum timestep size hadn't been limited to 0.1. The mesh in both cases consisted of twenty 9-noded elements. Furthermore, a one degree temperature range was used for the enthalpy method while for the source method the phase change was isothermal. The example was repeated using the source method with the maximum time step limited to one tenth of the previous value. The required time in this case was reached in 490 steps. The results are shown in Figure 5.7, which compare well with the accuracy of the enthalpy method. This finding does not agree with that of Dalhuijsen and Segal [15], who find the heat source (fictitious heat-flow) method inaccurate. The example was repeated once more using a 9-noded mesh of only nine elements as shown in Figure 5.8, using the source method. A very quick solution was obtained

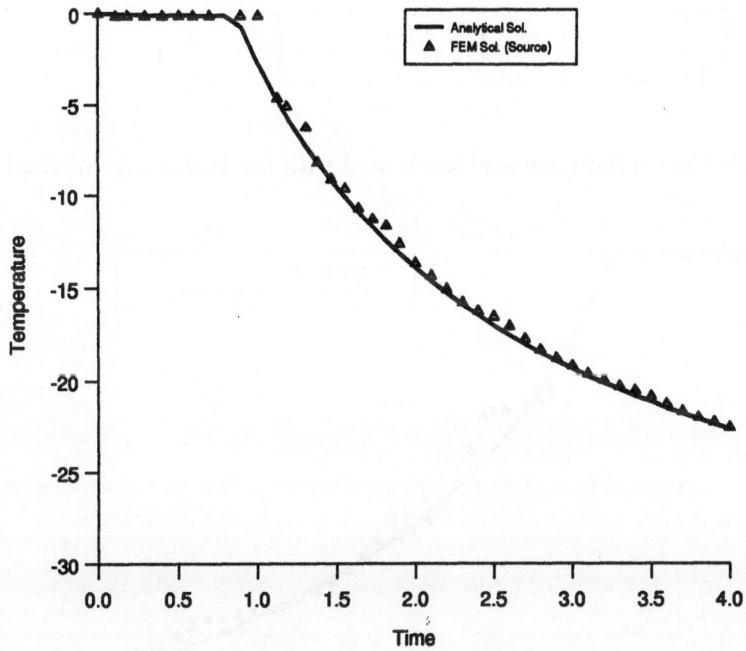

Figure 5.6: Source method vs the analytical solution.

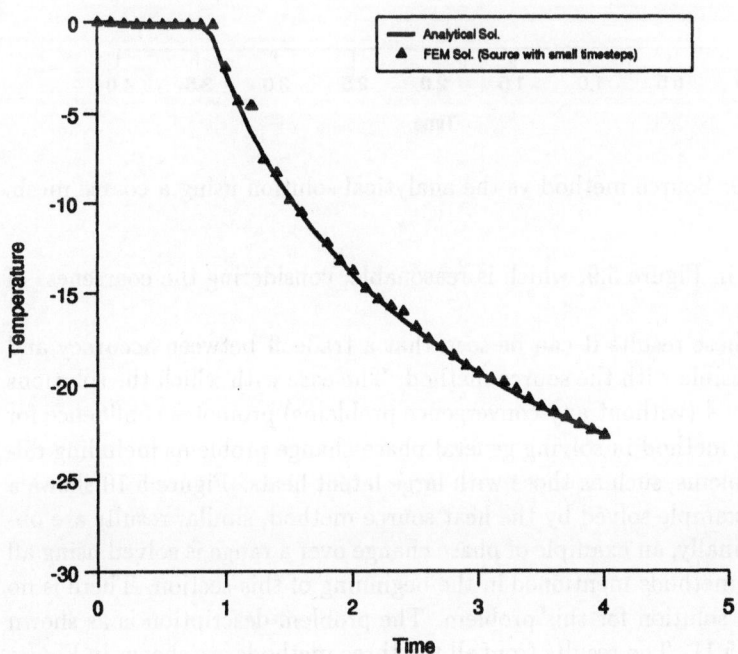

Figure 5.7: Source method vs the analytical solution using small timesteps.

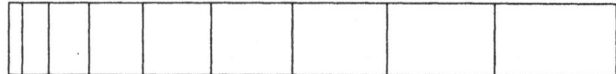

Figure 5.8: Coarse finite element mesh used with the heat source method

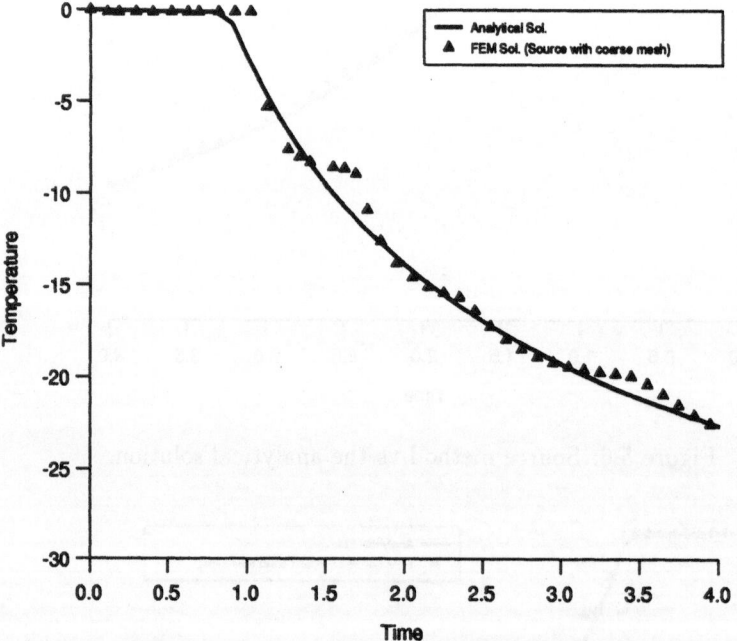

Figure 5.9: Source method vs the analytical solution using a coarse mesh.

as shown in Figure 5.9, which is reasonable, considering the coarseness of the mesh.

From these results it can be seen that a tradeoff between accuracy and cost is possible with the source method. The ease with which the solutions are obtained (without any convergence problems) promotes confidence for using this method in solving general phase change problems including difficult problems, such as those with large latent heats. Figure 5.10 shows a melting example solved by the heat source method, similar results are obtained. Finally, an example of phase change over a range is solved using all the three methods mentioned in the beginning of this section. There is no analytical solution for this problem. The problem description is as shown in Figure 5.11. The results from all the three methods are shown in Figure 5.12. The effective capacity method of [5] (which completelely failed to solve the previous problems with a 1° phase change interval) overestimates

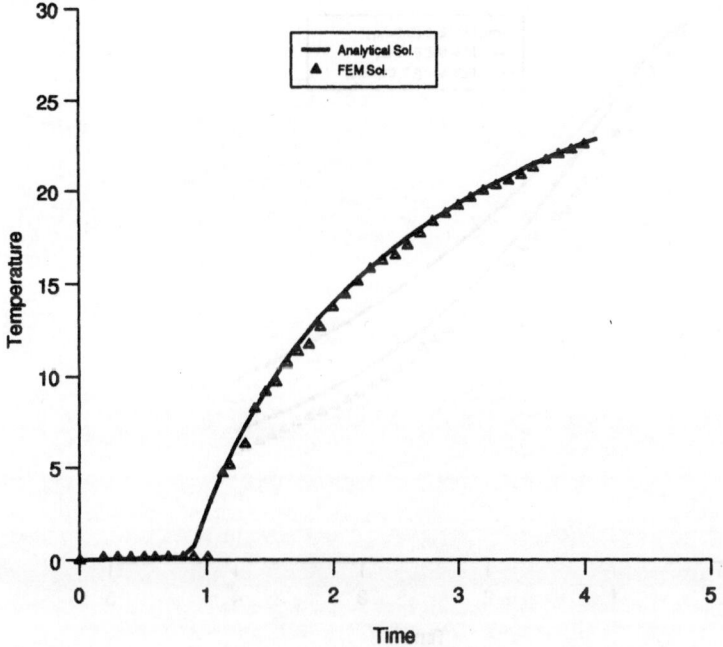

Figure 5.10: Source method vs the analytical solution for melting example.

$$L = 70.26$$

$$T = -45.0 \quad \boxed{T_o = 0.0 \quad T_s = -10.15 \quad T_l = -0.15} \quad \rho c = 1.0$$

$$x = 1.0 \qquad\qquad x = 4.0 \quad k = 1.0$$

Figure 5.11: Solidification over a temperature range example.

the temperature grossly. The enthalpy and the source methods roughly agree. The enthalpy method again needed roughly 1500 timesteps, while the source method needed only 70. The effective capacity method result was obtained from 120 timesteps.

5.5 Conclusion

It is concluded from the above discussion of the results of the benchmark example that the heat source method is the most efficient and flexible method available for analysing phase change problems in a fixed mesh finite element analysis. The one drawback of the source method seems to be the lack of smoothness of the temperature profiles for coarse meshes and large

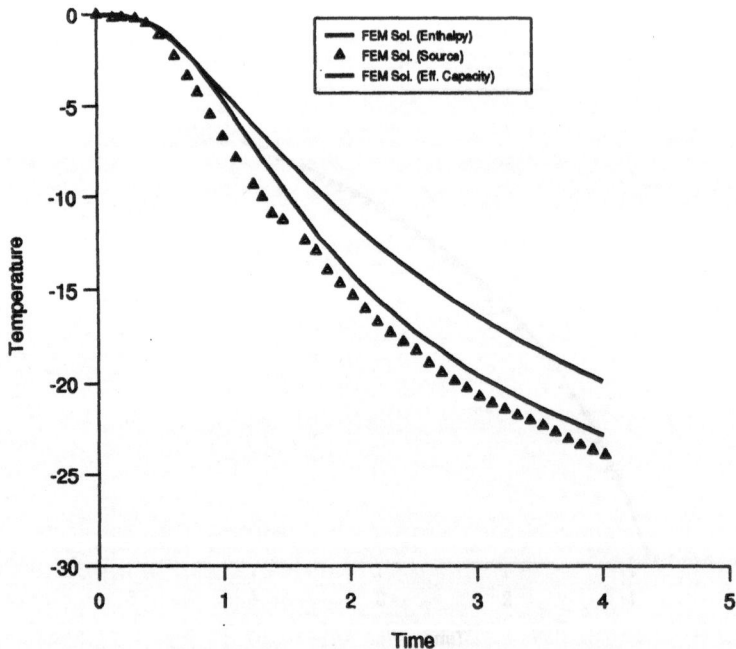

Figure 5.12: Solidification over a temperature range $(T_l - T_s = 10.0)$.

timesteps. This is due to the localised nature of the latent heat evolution, because of lumping all the latent heat at the nodes. This unsmoothness is relatively less noticable for problems with phase change over a range of temperature due to a more gradual latent heat release/absorption. The smoothness of the temperature profiles from the enthalpy method is due to the 'smearing' effect obtained by integrating the enthalpy over the whole element. The effective heat capacity method is only suitable for phase change over a very wide range of temperature.

References

[1] S.DelGuidice, G.Comini, and R.W.Lewis. Finite element simulation of freezing processes in soils. *International Journal of Numerical and Analytical Methods in Geomechanics*, 2:223–235, 1978.

[2] P.E.Frivik, E.Thorbergsen, S.DelGuidice, and G.Comini. Thermal design of pavement structures in seasonal frost areas. *Journal of Heat Transfer*, 99:533–540, 1977.

[3] H.C.Huang and R.W.Lewis. Adaptive analysis for heat flow problems using error estimation techniques. In *Sixth International Conference for Numerical Methods in Thermal Problems*, Swansea, U.K., July 1989. Pineridge Press, Swansea.

[4] J.Crank. *Free and Moving Boundary Problems*. Clarendon Press, Oxford, 1984.

[5] M.T.Samonds. *Finite Element Simulation of Solidification in Sand Mould and Gravity Die Castings - Ph.D. Thesis*. University of Wales, Swansea, 1985.

[6] R.J.Sarjant and M.R.Slack. Internal temperature distribution in the cooling and reheating of steel ingots. *Journal of Iron and Steel Institute*, 177:428–444, 1954.

[7] J.G.Henzel and J.Keverian. Predicting casting solidification patterns with a computer. *Foundry*, pages 50–53, 1964.

[8] J.G.Henzel and J.Keverian. Predicting casting solidification patterns in a steel valve casting by means of a digital computer. *Metals Eng. Quarterly*, pages 39–44, 1965.

[9] J.G.Henzel and J.Keverian. Comparison of calculated and measured solidification patterns in a variety of steel castings. *AFS Cast Metals Research Journal*, pages 19–36, 1965.

[10] R.Weatherwax and O.K.Riegger. Computer-aided solidification study of a die cast aluminium piston. *AFS Transactions*, 85:317–322, 1977.

[11] P.N.Hansen. Numerical simulation of the solidification process. In *Solidification and Casting of Metals*, Sheffield, U.K., 1979. The Metals Society. Proceedings of the Conference.

[12] J.I.Soliman and E.A.Fakhroo. Finite element solution of heat transmission in steel ingots. *Journal of Mechanical Eng. Science*, 14:19–24, 1972.

[13] K.Morgan, R.W.Lewis, and K.N.Seetharamu. Modelling of heat flow and thermal stress in ingot casting. *Simulation*, pages 55–63, 1981.

[14] R.W.Lewis, K.N.Seetharamu, and K.Morgan. Application of the finite element method in the study of ingot castings. In *Solidification Technology in the Foundry and Cast House*, pages 40–43, Coventry, U.K., September 1983. The Metals Society. Proceedings of the Conference.

[15] A.J.Dalhuijsen and A.Segal. Comparison of finite element techniques for solidification problems. *International Journal for Numerical Methods in Engineering*, 23:1807–1829, 1986.

[16] M.Salcudean and Z.Abdullah. On the numerical modelling of heat transfer during solidification processes. *International Journal for Numerical Methods in Engineering*, 25:445–473, 1988.

[17] V.R.Voller, C.R.Swaminathan, and B.G.Thomas. Fixed grid techniques for phase change problems: A review. *International Journal for Numerical Methods in Engineering*, 30:875–898, 1990. Special thermal issue.

[18] J.Crank. How to deal with moving boundaries in thermal problems. In R.W.Lewis, K.Morgan, and O.C.Zienkiewicz, editors, *Numerical Methods in Heat Trafsfer*. Wiley, Chichester, 1981.

[19] K.O'Neill and D.R.Lynch. A finite element solution of freezing problems using a continuously deforming coordinate system. In R.W.Lewis, K.Morgan, and O.C.Zienkiewicz, editors, *Numerical Methods in Heat Trafsfer*. Wiley, Chichester, 1981.

[20] J. Yoo and B.Rubinsky. Numerical computation using finite elements for the moving interface in heat transfer problems with phase transformation. *Numerical Heat Transfer*, 6:209–222, 1983.

[21] R.W.Lewis K.Morgan and O.C.Zienkiewicz. An improved algorithm for heat conduction problems with phase change. *International Journal for Numerical Methods in Engineering*, 12:1191–1195, 1978.

[22] R.W.Lewis and H.C.Huang. A finite element analysis of fire resisting cabinets using an adaptive remeshing technique. *Applied Mathematical Modelling*, (to be published in 1991).

[23] R.W.Lewis and P.M.Roberts. Finite element simulation of solidification problems. *Applied Scientific Research*, 44:61–92, 1987.

[24] E.C.Lemmon. Multidimensional integral phase change approximations for finite element conduction codes. In R.W.Lewis, K.Morgan, and O.C.Zienkiewicz, editors, *Numerical Methods in Heat Trafsfer*. Wiley, Chichester, 1981.

[25] R.Viskanta. Phase change heat transfer. In G.A.Lane, editor, *Solar Heat Storage Latent Heat Materials*. CRC Press, 1983.

[26] V.R.Voller, M.Cross, and N.C.Markatos. An enthalpy method for convection/diffusion phase change. *International Journal for Numerical Methods in Engineering*, 24:271–284, 1987.

[27] W.D.Rolph and K.J.Bathe. An efficient algorithm for analysis of nonlinear heat transfer with phase changes. *International Journal for Numerical Methods in Engineering*, 18:119–134, 1982.

[28] J.Roose and O.Storrer. Modelization of phase changes by fictitious heat flow. *International Journal for Numerical Methods in Engineering*, 20:217–225, 1984.

[29] O.C.Zienkiewicz. *The Finite Element Method*. McGraw-Hill Book Company (UK) Limited, London, 1977.

[30] H.S.Carslaw and J.C.Jaeger. *Conduction of Heat in Solids*. Clarendon Press, Oxford, 1959.

[31] Jr. C.Hastings. *Approximations for Digital Computers*. Princeton University Press, Princeton,N.J., 1955.

[26] .... 1984.

[27] J.A. Wood, 1984 and A.J. Marshall, An exact method for treating .... please change in running journal .... Methods in Engineering, vol.11, 234, 1997.

[28] W.D.Collins, J.K.J Rao, An efficient algorithm for analysis of non-linear heat transfer with phase changes, International Journal for Numerical Methods in Engineering, 19:119–121, 1979.

[29] J. Crank and R.S. Gupta, Modelization of phase changes by a finite element .... latent .... , International Journal for Numerical Methods in Engineering, vol. 30(24) 32, 1990.

[30] O.C.Zienkiewicz, The Finite Element Method, McGraw Hill Book Company (UK) Limited, London, 1977.

[31] H.S. Carslaw and J.C. Jaeger, Conduction of Heat in Solids, Clarendon Press, Oxford, 1959.

[32] R.C. Buckius, Approximation for Digital Computers, Princeton University Press, Princeton N.J., 1955.

Chapter 6

Adaptive Heat Transfer Analysis

6.1 Introduction

Techniques for error estimation have recently been developed for stress analysis problems. These methods may be considered as optimizing the finite element analysis according to the intrinsic behaviour of the given problem. This type of error estimation is now applied to heat flow analyses. The technique will first be demonstrated for steady state thermal problems and then extended to transient problems.

As mentioned in the previous chapters, to solve the heat transfer problems by the finite element method the domain of interest is first subdivided into a number of discrete subdomains or elements. Variational, Galerkin or other techniques can be used to obtain an integral formulation, which is usually referred to as the 'weak' form. After evaluation of the integrals a set of algebraic equations, in matrix form, is obtained for each element. Once such equations are formulated for all the elements, the global matrices are then assembled and the final solution is obtained.

The size and placement of these elements largely determines the accuracy with which the problem can be solved. Reducing the element size and thereby increasing the number of nodal points usually yields a more accurate solution but at the cost of an increased CPU time and memory requirement. The key to the efficient and economic solution of problems is not merely the *number* of nodal points and elements but also their *placement*. Regions with large gradients (e.g. a singularity, stress concentration or region of high heat flux) will need a high mesh density, with quiescent regions requiring a comparatively coarser mesh. In many real engineering situations, it is desirable either to obtain the most accurate solution possible within an upper problem size limit, or to try to capture a numerically awkward feature without using an excessive number of elements.

The premise of adaptive procedures is that, by making use of the mathematics of error analysis, a finite element program can determine which region needs refining and automatically adapt the mesh to suit the problem. The error is calculated in each element and is compared with a predefined limit. It is expected that for a given accuracy of the solution each element has approximately the same level of error. Therefore, any element with an error above, and in some cases below, must be adjusted to match this error level. The process is repeated, if necessary, with the ultimate aim that every element contains the same predefined, allowable error, thus yielding an optimal mesh.

There are two distinct phases of the adaptive solution process: error analysis and mesh refinement. Error analysis allows us to calculate for each element, in the case of error estimation, an absolute value for the error, and in the case of error indication, an element value relative to other elements in the mesh. Much of the early mathematical work in error analysis was due to Babuska and Rheinboldt [1, 2], but was drawn together, in the form of an error estimate, in a notable paper by Kelly *et al.* [3]. A companion paper by Gago *et al.* [4] was later published which offered several strategies for using an error estimator to refine a mesh.

Once an estimate of the error in each element has been obtained, there are several ways of going about the next phase of the adaptive process, i.e. improving the mesh. The two basic approaches are *p — refinement* and *h — refinement*. *p*-refinement involves increasing the order of the polynomial approximation within an element while *h*-refinement simply means reducing the subdivision size. *p*—refinement, especially when combined with a hierarchical formulation, has several advantages–it is more efficient, converges faster and appeals to the purist by virtue of its mathematical elegance. However, incorporating *p*—refinement generally means restructuring, if not rewriting, an existing finite element code. *h*—refinement is more universally accepted [5, 6] and has been employed in the content of this chapter.

Having decided on adopting the *h*—refinement method there is another choice to be made, *element subdivision* or *mesh regeneration*. With the element subdivision approach, every element that exceeds the allowable error threshold is subdivided into smaller elements. This is most effective when using four-node elements as it is otherwise very difficult to achieve the desired density distribution. However, constrained nodes are introduced which must be dealt with, and the method allows only one level of subdivision at a time.

The remeshing approach involves completely regenerating the mesh, either in regions of high error only, or over the entire domain. The advantage of regenerating the entire mesh is that areas can be coarsened if the calculated error is below the allowable error. It is this that allows the generation of a truly optimal mesh in which every element has approximately the same, predefined, level of error, and it was this fact that led the authors to use the mesh regeneration approach in this work. One of the disadvantages of mesh regeneration is that a high degree of spatial flexibility is required if use is to be made of the abundant information provided by an error estimation procedure. For planar problems we can automatically generate both triangles and quadrilaterals in regions of arbitrary geometry using a mesh generator based on the Delaunay triangulation procedure. The details of adaptive heat transfer analysis is presented as follows.

6.2 Error Estimation for Heat Conduction

The governing field equation for steady-state heat conduction is derived in Chapter 2 as

$$\frac{\partial}{\partial x}(k_x \frac{\partial T}{\partial x}) + \frac{\partial}{\partial y}(k_y \frac{\partial T}{\partial y}) + Q = 0 \tag{6.1}$$

with the boundary conditions of

$$T = T_0 \quad \text{on} \quad \Gamma_T \tag{6.2}$$

and

$$-k\frac{\partial T}{\partial n} = \bar{q} \quad \text{on} \quad \Gamma_q \tag{6.3}$$

Discretizing in space by the finite element method yields a system of algebraic equations. The temperature field is then obtained by solving this set of equations.

As mentioned above, the h-refinement technique and error estimation algorithm is adopted for heat flow problems. Errors of interest in the heat flow problem are given by the following difference expressions viz:

$$\mathbf{e}_T = \mathbf{T} - \hat{\mathbf{T}} \tag{6.4}$$

or

$$\mathbf{e}_q = \mathbf{q} - \hat{\mathbf{q}} \tag{6.5}$$

with

$$q_x = -k_x \frac{\partial T}{\partial x} \tag{6.6}$$

$$q_y = -k_y \frac{\partial T}{\partial y} \tag{6.7}$$

where $\mathbf{T}$ and $\mathbf{q}$ are actual temperature and heat flux, while $\hat{\mathbf{T}}$ and $\hat{\mathbf{q}}$ are the solution obtained from the finite element analysis.

However, it is more reasonable to use a scalar or a norm to express these errors. The error norm in the heat flow problem can be defined as follows

$$||e|| = \left[\int_\Omega k_x (\frac{\partial T}{\partial x} - \frac{\partial \hat{T}}{\partial x})^2 + k_y (\frac{\partial T}{\partial y} - \frac{\partial \hat{T}}{\partial y})^2 dx dy \right]^{1/2} \tag{6.8}$$

or in the vector form as

$$||e|| = \left[\int_{\Omega} (\nabla \mathbf{T} - \nabla \hat{\mathbf{T}})^T \mathbf{k} (\nabla \mathbf{T} - \nabla \hat{\mathbf{T}}) d\Omega \right]^{1/2} \tag{6.9}$$

where

$$\mathbf{k} = \begin{bmatrix} k_x & 0 \\ 0 & k_y \end{bmatrix} \tag{6.10}$$

It can be shown that [7]

$$\int_{\Omega} (\nabla \mathbf{T})^T \mathbf{k} \nabla \hat{\mathbf{T}} d\Omega = \int_{\Omega} (\nabla \hat{\mathbf{T}})^T \mathbf{k} \nabla \mathbf{T} d\Omega = \int_{\Omega} (\nabla \hat{\mathbf{T}})^T \mathbf{k} \nabla \hat{\mathbf{T}} d\Omega \tag{6.11}$$

Thus, we have

$$||e||^2 = \int_{\Omega} (\nabla \mathbf{T})^T \mathbf{k} \nabla \mathbf{T} d\Omega - \int_{\Omega} (\nabla \hat{\mathbf{T}})^T \mathbf{k} \nabla \hat{\mathbf{T}} d\Omega \tag{6.12}$$

if we define

$$||q||^2 = \int_{\Omega} (\nabla \mathbf{T})^T \mathbf{k} \nabla \mathbf{T} d\Omega$$

$$||\hat{q}||^2 = \int_{\Omega} (\nabla \hat{\mathbf{T}})^T \mathbf{k} \nabla \hat{\mathbf{T}} d\Omega \tag{6.13}$$

Equation (6.12) can be rewritten as

$$||e||^2 = ||q||^2 - ||\hat{q}||^2 \tag{6.14}$$

Here the term $||q||$ may be considered as the total heat flow dissipation in the whole domain. Such a definition allows a practical representation of the error norm in terms of a percentage error, that is

$$\eta = \frac{||e||}{||q||} \times 100\% \tag{6.15}$$

The significance of the term $||e||/||q||$, can be realized from the following analysis. We consider a special boundary condition where $T = 0$ and $\frac{\partial T}{\partial n} = 0$ on the domain boundary Γ. Using Green's lemma [8] we can write

$$||q||^2 = \int_{\Omega} (\nabla \mathbf{T})^T \mathbf{k} \nabla \mathbf{T} d\Omega = - \int_{\Omega} \mathbf{T} \nabla (\mathbf{k} \nabla \mathbf{T}) d\Omega \tag{6.16}$$

and similarly for $||\hat{q}||^2$

$$||\hat{q}||^2 = \int_{\Omega} (\nabla \hat{\mathbf{T}})^T \mathbf{k} \nabla \hat{\mathbf{T}} d\Omega = \int_{\Omega} (\nabla \hat{\mathbf{T}})^T \mathbf{k} \nabla \mathbf{T} d\Omega = - \int_{\Omega} \hat{\mathbf{T}} \nabla (\mathbf{k} \nabla \mathbf{T}) d\Omega \tag{6.17}$$

Thus, from Equations (6.16) and (6.17) we have

$$||e||^2 = ||q||^2 - ||\hat{q}||^2 = - \int_{\Omega} (\mathbf{T} - \hat{\mathbf{T}}) \nabla (\mathbf{k} \nabla \mathbf{T}) d\Omega \tag{6.18}$$

from the mean value theorem it is found that

$$\eta^2 = \frac{||e||^2}{||q||^2} = \left| \frac{(\mathbf{T} - \hat{\mathbf{T}})^*}{\mathbf{T}^*} \right| \tag{6.19}$$

where $(\mathbf{T} - \hat{\mathbf{T}})^*$ and $\mathbf{T}^*$ are temperatures at appropriate points. Since, a uniform distribution of errors is expected, therefore η^2 can be thought of as representing an average error for the temperature field. It is evident that the temperature errors will be much lower than those of the gradient. For instance, an error of 20 per cent in the gradient values should roughly correspond to an error of 4 per cent in the temperature.

6.3 Higher Order Approximation

Before the implementation of the adaptive procedure is discussed the problem of calculating the error norm $||e||$ must be addressed as it contains so-called 'exact' values of the temperature gradients, which are obviously not available. It is suggested by Zienkiewicz and Zhu that a globally smoothed value may be taken as an approximation of higher order to that given by the finite element solution [9]. In this smoothing process, it is assumed that the approximation quantities are interpolated by the same basis functions as the variable T and that they fit the original ones in a least square sense. This process is known as *recovery*. In this study the smoothed continuous values of temperature gradients are represented in the following manner as given by Hinton and Campell [10]. However, basis functions of the same order have been used here for smoothing as for the main variable T, unlike lower order basis function used in [10]. The smoothing procedure involves representing the smoothed gradients G in the element in terms of their unknown nodal values $\tilde{G}_i$, i.e.

$$\tilde{G} = \sum_{i=1}^{n} N_i \tilde{G}_i \tag{6.20}$$

where n is the number of nodes per element and N_i are smoothing shape functions which are usually defined on the original mesh of elements. If $\hat{G}$ represents the non-smoothed values obtained from the original finite element analysis, then the problem becomes one of finding the nodal values $\tilde{G}_i$ which minimize the functional

$$I = \int_\Omega (\tilde{G} - \hat{G})^2 d\Omega \tag{6.21}$$

Hence for I to be a minimum

$$\frac{\partial I}{\partial \tilde{G}_i} = 0 \tag{6.22}$$

This leads to a set of linear simultaneous equations in $\tilde{G}_i$ as follows

$$\left(\int_\Omega N_i N_j d\Omega \right) \tilde{G}_i = \int_\Omega N_i \hat{G} d\Omega \qquad (6.23)$$

It may be re-written as

$$\sum M_{ij} \tilde{G}_i = f_i \qquad (6.24)$$

The 'smoothing matrix' M_{ij} in each element is given as:

$$M_{ij}^e = \int_{-1}^1 \int_{-1}^1 N_i N_j det\mathbf{J} d\xi d\eta \qquad (6.25)$$

and the 'smoothing forces' f_i in each element are given as:

$$f_i^e = \int_{-1}^1 \int_{-1}^1 N_i \hat{G} det\mathbf{J} d\xi d\eta \qquad (6.26)$$

When Gauss-Legendre integration rule is applied to the 'smoothing forces', we obtain

$$f_i^e = \sum N_i(\xi_k, \eta_k) det\mathbf{J}(\xi_k, \eta_k) \hat{G}(\xi_k, \eta_k) \qquad (6.27)$$

where $\hat{G}(\xi_k, \eta_k)$ are the temperature gradients at Gauss points (ξ_k, η_k) which are available from the finite element solutions. It has been proved that the temperature obtained by the finite element method is most accurate at nodal points [11], whereas the temperature gradient is most accurate at Gauss points [12]. This characteristic is often referred to as superconvergence phenomena. After solving this equation system, globally smoothed values of temperature gradients are obtained with the higher approximation to the exact solution. However, for quadratic elements, it has been shown that [13] a local smoothing procedure produces even better gradient values at nodal points. In this method, local gradient functions are generated by a least square fit from values at Gauss points in adjacent elements.

With a higher approximation available, it is now possible to carry on with adaptive analysis using error estimation.

6.4   Implementation of the Adaptive Procedure

The first step for implementing adaptive analysis is to specify a maximum permissible error $\bar{\eta}$ that is to be achieved at the end of the analysis. The requirement for a near optimum analysis is that all the elements of the final mesh must contain an approximately equal error. After the first analysis, from a preliminary coarse mesh, the square of the total error is calculated, which is the sum of all individual element contributions, that is

$$||q||^2 \approx ||\hat{q}||^2 = \sum_{e=1}^m \int_{\Omega_e} (\nabla \hat{\mathbf{T}})^T \mathbf{k} \nabla \hat{\mathbf{T}} d\Omega_e \qquad (6.28)$$

or more explicitly, for two dimensions

$$||q||^2 \approx ||\hat{q}||^2 = \sum_{e=1}^{m} \int_{\Omega_e} \left(k_x (\frac{\partial \hat{T}}{\partial x})^2 + k_y (\frac{\partial \hat{T}}{\partial y})^2 \right) dx dy \qquad (6.29)$$

where m is the total number of elements. Now, the maximum permissible error for each element can be calculated by distributing $||q||^2$ equally over all the elements, i.e.

$$||\bar{e}||_e^2 \leq \bar{\eta} \left(\frac{||q||^2}{m} \right) \qquad (6.30)$$

where $\bar{\eta}$ is some specified maximum value. The approximate error in each element after the first analysis can be calculated according to Equation (6.8) using, of course, the smoothed values of temperature gradients instead of the exact values. This can be written for a two-dimensional element e as

$$||e||_e = \left[\int_{\Omega_e} k_x (\frac{\partial \tilde{T}}{\partial x} - \frac{\partial \hat{T}}{\partial x})^2 + k_y (\frac{\partial \tilde{T}}{\partial y} - \frac{\partial \hat{T}}{\partial y})^2 dx dy \right]^{1/2} \qquad (6.31)$$

where $\frac{\partial \tilde{T}}{\partial x}, \frac{\partial \tilde{T}}{\partial y}$ are the smoothed values of temperature gradients and $\frac{\partial \hat{T}}{\partial x}, \frac{\partial \hat{T}}{\partial y}$ are the finite element solutions.

Now this error is compared for all elements to the maximum permissible error in an element as calculated above and used to modify the mesh for a second analysis. If we define a variable ξ_e, where

$$\xi_e = \frac{||e||_e}{||\bar{e}||_e} \qquad (6.32)$$

If $\xi_e > 1$, the size of element e must be reduced and the mesh will require refinement, otherwise, the size of the element must be increased and the mesh will be coarsened. Thus the predicted size of the new element can be calculated from the current element size as follows

$$\bar{h}_e = \frac{h_e}{\xi_e^{1/P}} \qquad (6.33)$$

where $\bar{h}_e$ is the predicted element size, h_e is current element size and P is the order of the shape functions.

6.5 Steady State Benchmark Example

In the numerical example given, both linear and quadratic triangular elements are employed for the adaptive remeshing technique [8, 14].

The first example to test the adaptive procedure, as outlined above, was obtained from a list of selected benchmarks published by NAFEMS [15]. A rectangular plate shown in Figure 6.1 is subjected to three different boundary conditions: 1. An insulated boundary along the edge AD, 2. A given temperature of 100°C at the edge AB, 3. A surface convection to ambient temperature of 0°C along the edge BC and DC. It is noted that the severe discontinuity in boundary conditions at point B gives rise to a quasi-singular type behaviour at this location. This fact, combined with the relatively coarse meshes specified in [15], made it almost impossible to obtain a good comparison to the target value of temperature at point E which lies at a distance of 0.2m above B on the edge BC. The analytical solution for the temperature at this point is 18.2535°C.

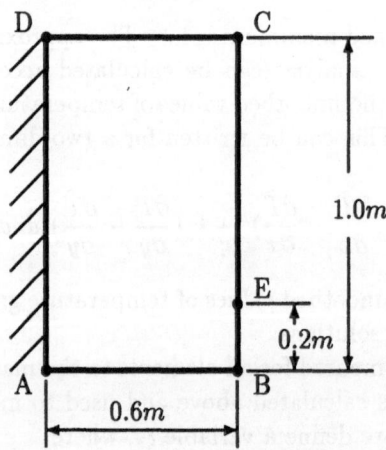

Figure 6.1: Two dimensional heat transfer with convection

The analytical solution obtained from solving the differential equation with the specified boundary conditions is derived in [16] and is given by

$$T = 2hT_{AB} \sum_{n=1}^{\infty} \frac{cos\alpha_n x[\alpha_n cosh\alpha_n(b-y) + \underline{h}sinh\alpha_n(b-y)]}{[(\alpha_n^2 + \underline{h}^2)a + \underline{h}](\alpha_n cosh\alpha_n b + \underline{h}sinh\alpha_n b)cos\alpha_n a} \quad (6.34)$$

where

$$
\begin{aligned}
a &= 0.6m \\
b &= 1.0m \\
\underline{h} &= h/k \\
k &= \text{thermal conductivity} = 52 \ W/m°C \\
h &= \text{heat transfer coefficient} = 750 \ W/m°C \\
T_{AB} &= 100°C
\end{aligned}
$$

and α_n are the roots of "$\alpha \ tan \ \alpha a = \underline{h}$" which may be solved using the iterative method as follows.

$$\alpha_n = \frac{1}{a}[\tan^{-1}(\underline{h}/\alpha_n) + (n-1)\pi]$$

Table 6.1: The first ten roots of "$\alpha \tan \alpha a = \underset{\sim}{h}$"

n	α_n
1	2.3489
2	7.0923
3	11.9377
4	16.8861
5	21.9141
6	26.9977
7	32.1194
8	37.2674
9	42.4340
10	47.6141

where $n = 1 \longrightarrow \infty$. The first ten roots are listed in the Table 6.1.

This analytical solution is shown for the whole domain in Figure 6.2, from which it is clearly seen that the steep temperature gradient between points B and E requires the use of a relatively fine mesh in this area. In many practical problems it is not immediately obvious as to where a fine mesh may be required and this is where the advantage of the adaptive finite element method lies.

When a uniform mesh of thirty triangular elements is used, which was recommended in reference [15], a temperature error of −24.6% was obtained for the three noded elements, whereas an error of −1.6% was obtained for six noded elements at the point E.

Now, the adaptive error estimation technique is applied to this same problem. For both linear and quadratic triangular elements a uniform mesh, similar to that used in reference [15], is adopted as the initial mesh (Figure 6.3). A target error of 10% is prescribed for the linear element. The mesh shown in Figure 6.4a was automatically generated according to the estimated errors and indicates a norm error of 14.3%, which implies an average temperature error of 2.0%. However, the error at the point E compared to the analytical solution was less than 0.6%. After further remeshing (Figure 6.4b), a norm error of 9.8% was indicated. The mesh can be seen to be finer in the area of points B and E, as expected, and much coarser elsewhere.

When six noded triangular elements are employed with a target error of 5% a much better result was obtained. The two subsequent meshes in Figure 6.5a-b were generated according to the adaptive procedure which leads to a norm error of 2.8%. This corresponds to a temperature error of approximately 0.08%. The solution obtained from this mesh is virtually indistinguishable from the analytical solution.

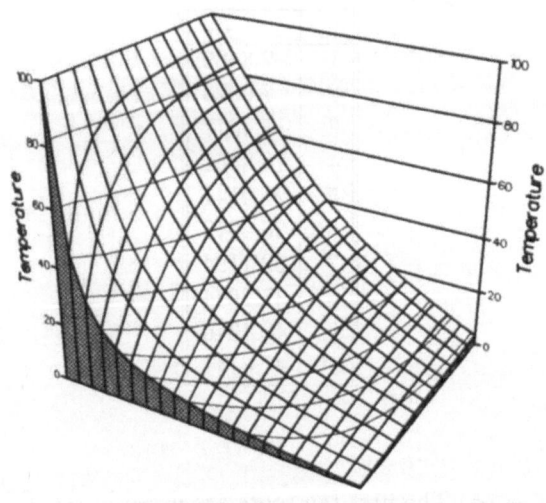

Figure 6.2: Temperature distribution of analytical solution

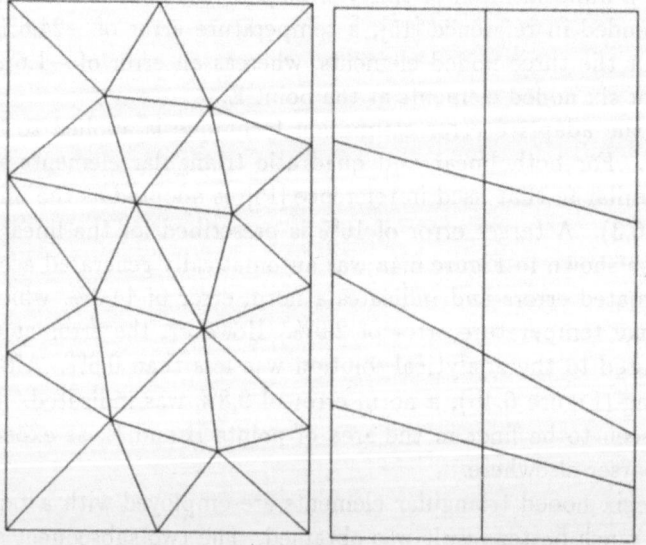

Figure 6.3: Initial meshes for triangular and quadrilateral elements

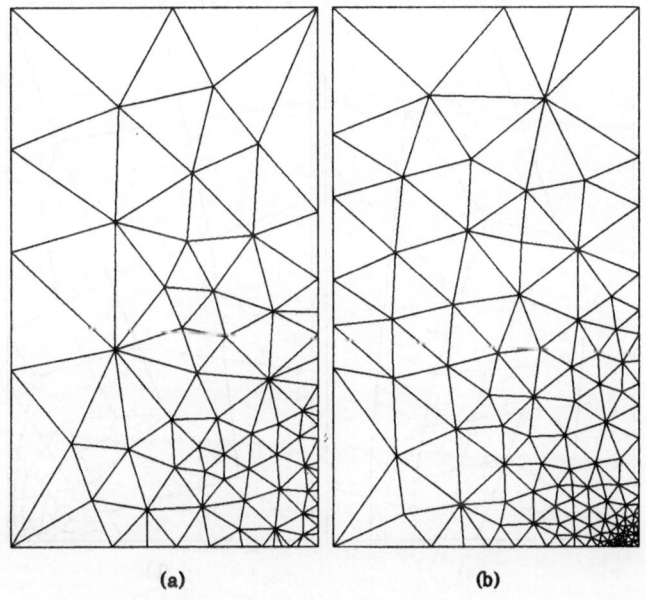

Figure 6.4: Set of refined meshes (3 node) based on the smoothed solution

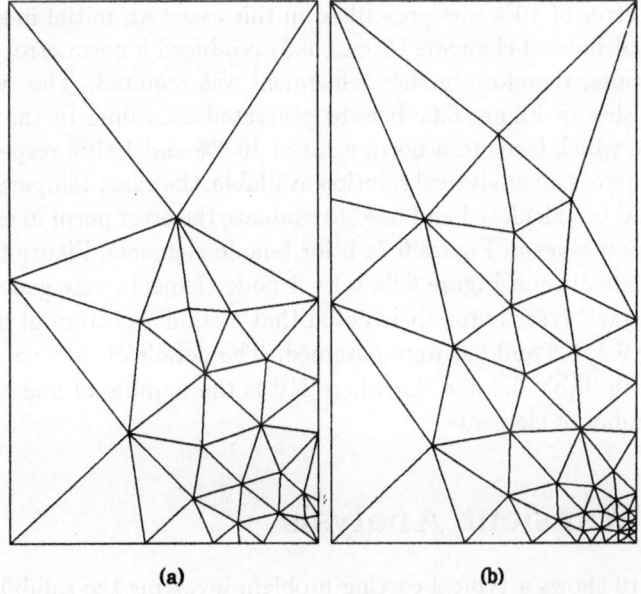

Figure 6.5: Set of refined meshes (6 node) based on the smoothed solution

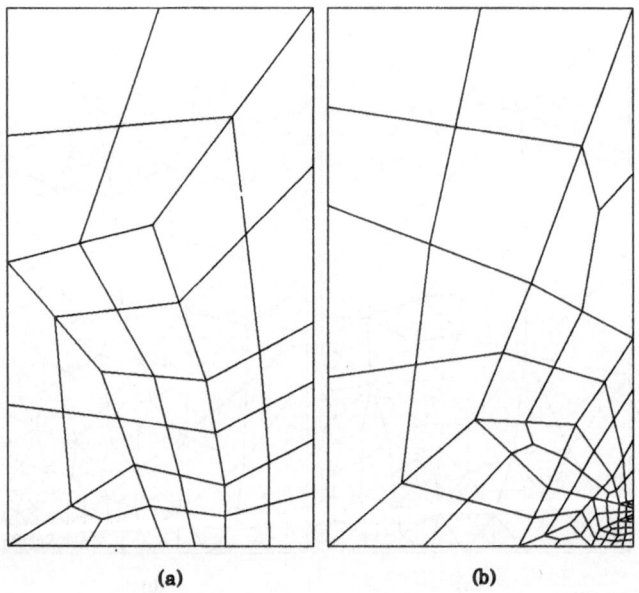

Figure 6.6: Set of refined meshes (4 node) based on the smoothed solution

Four noded quadrilateral elements are used to solve the same problem. A target error of 10% was prescribed in this case. An initial mesh with 8 linear quadrilateral elements (Figure 6.3) produced a norm error of 21.6% in the results, therefore further refinement was required. The two subsequent meshes in Figure 6.6a-b were generated according to the adaptive procedure which leads to a norm error of 16.0% and 10.0% respectively.

Since there is an analytical solution available, the exact temperature gradient could be calculated and used to evaluate the error norm in expression (6.8). The meshes in Figure 6.7a-b for 3-node elements, Figure 6.8a-b for 6-node elements and Figure 6.9a-b for 4-node elements were generated using the 'exact' error norm. It is noted that a similar pattern of meshes as in Figure 6.4, 6.5 and 6.6 were obtained. The details of such comparisons are given in Table 6.2 and 6.3 where NP is the number of nodes and NE is the number of elements.

6.6   Transient Analysis

Figure 6.10 shows a typical casting problem involving the solidification of liquid metal poured into a sand mould. A thin metal fin protrudes from the main body of the casting into the mould. The initial temperature of the metal is $700°C$ and that of the sand is $20°C$. The latent heat

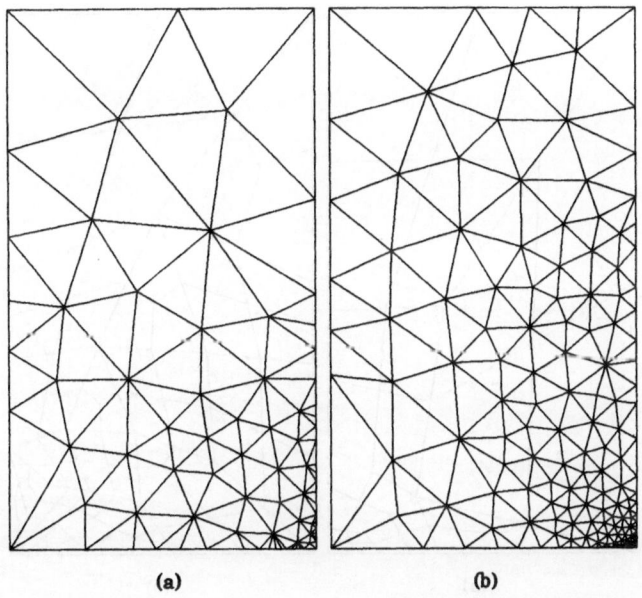

Figure 6.7: Set of refined meshes (3 node) based on the exact solution

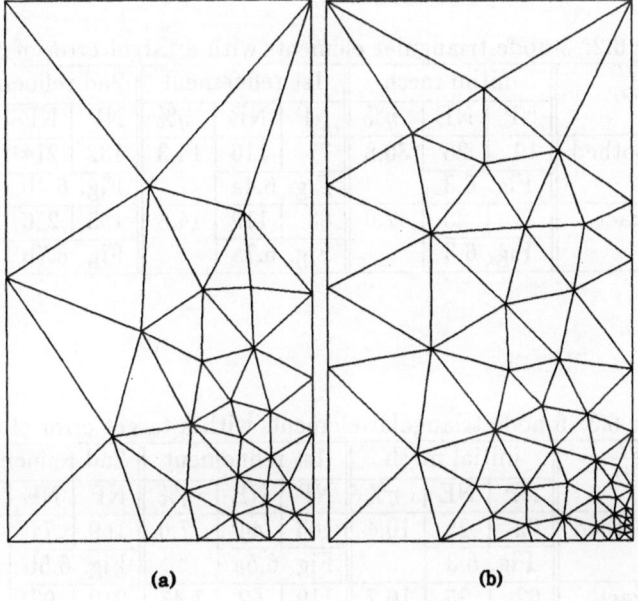

Figure 6.8: Set of refined meshes (6 node) based on the exact solution

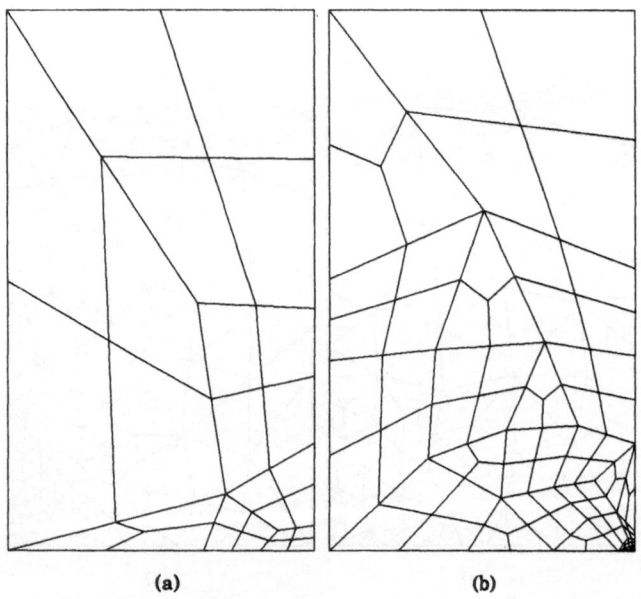

(a)   (b)

Figure 6.9: Set of refined meshes (4 node) based on the exact solution

Table 6.2: 3-node triangular elements with a target error of 10.0%

$\frac{\partial T}{\partial x_i}$	initial mesh			1st refinement			2nd refinement		
	NP	NE	$\eta\%$	NP	NE	$\eta\%$	NP	NE	$\eta\%$
smoothed	19	25	30.8	77	116	14.3	132	214	9.8
	Fig. 6.3			Fig. 6.4a			Fig. 6.4b		
exact	19	25	42.0	83	119	14.8	155	256	9.6
	Fig. 6.3			Fig. 6.7a			Fig. 6.7b		

Table 6.3: 6-node triangular elements with a target error of 5.0%

$\frac{\partial T}{\partial x_i}$	initial mesh			1st refinement			2nd refinement		
	NP	NE	$\eta\%$	NP	NE	$\eta\%$	NP	NE	$\eta\%$
smoothed	62	25	10.4	103	44	7.0	169	74	2.8
	Fig. 6.3			Fig. 6.5a			Fig. 6.5b		
exact	62	25	16.7	119	52	7.47	212	93	3.6
	Fig. 6.3			Fig. 6.8a			Fig. 6.8b		

Table 6.4: 4-node quadrilateral elements with a target error of 10.0%

$\frac{\partial T}{\partial x_i}$	initial mesh			1st refinement			2nd refinement		
	NP	NE	$\eta\%$	NP	NE	$\eta\%$	NP	NE	$\eta\%$
smoothed	15	8	21.6	45	35	16.0	93	77	10.0
	Fig. 6.3			Fig. 6.6a			Fig. 6.6b		
exact	15	8	27.9	37	27	16.3	105	89	10.2
	Fig. 6.3			Fig. 6.9a			Fig. 6.9b		

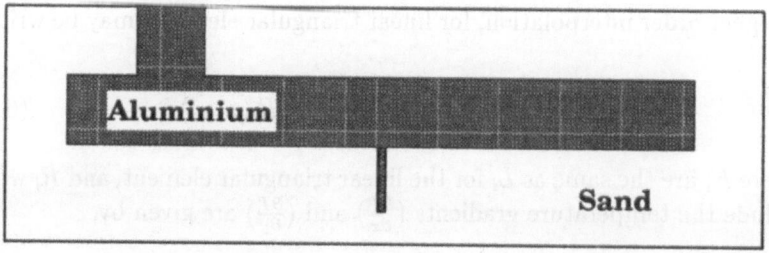

Figure 6.10: Aluminium casting in a sand mould

of solidification for the metal is accounted for by the enthalpy method discussed in Chapter 4. This configuration, though not very complicated, does not immediately suggest the type of mesh that must be used for a reasonable solution. For this particular problem the choice of a good mesh is further complicated by the fact that a solidification front exists which moves as the transient solution evolves and latent heat is released. An adaptive remeshing procedure is well suited to such problems.

As this is a transient problem the adaptive procedure is varied slightly. The approach preferred is to check the error after every timestep or a fixed number of timesteps and modify the mesh accordingly. The additional

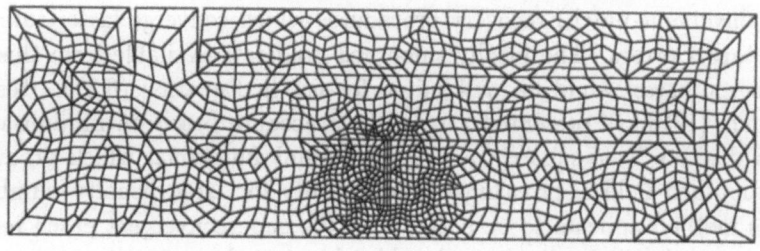

Figure 6.11: The initial finite element mesh

work involved here is the need to interpolate the nodal information from the previous mesh to the new mesh in order to continue the time integration process. The interpolation for linear triangular elements may be conducted as follows

$$T(x,y) = L_1(x,y)T_1 + L_2(x,y)T_2 + L_3(x,y)T_3 \qquad (6.35)$$

where L_1, L_2, L_3 are area coordinates at (x,y) and T_1, T_2, T_3 are temperatures at the element nodes. Such linear interpolation is not satisfactory for most problems. An interpolation suggested by Sampaio *et. al.* [17] uses the smoothed gradients available from the error estimation process. A higher order interpolation, for linear triangular elements may be written as,

$$T(x,y) = \sum_{i=1}^{3} N_i(x,y)\left(T_i + R_i(x,y)\right) \qquad (6.36)$$

where N_i are the same as L_i for the linear triangular element, and R_i which include the temperature gradients $\left(\frac{\partial T}{\partial x}\right)$ and $\left(\frac{\partial T}{\partial y}\right)$ are given by,

$$R_i(x,y) = 0.5\left((x-x_i)\left(\frac{\partial T}{\partial x}\right)_i + (y-y_i)\left(\frac{\partial T}{\partial y}\right)_i\right) \qquad (6.37)$$

This interpolation is equivalent to the the one given in reference [17], however it is cheaper to calculate as it does not contain second order terms. The globally smoothed temperature gradients available from the error estimation procedure could be used in the above formula, however, we have found that these produce spatial oscillations in the interpolated solution if the gradients are too steep (such as in the transient casting problem here). Using a lumped version of the global smoothing matrix of Equation (6.24) alleviates the problem of spatial oscillations. For the program **HADAPT** of Appendix B, we have employed the equivalent procedure of averaging the constant gradients (from a linear triangle) at the nodes, to obtain the smooth gradients for Equation (6.37). For the purpose of interpolation we convert other elements to the linear triangle in **HADAPT**.

Later works of Zienkiewicz and Zhu, such as [18], have used local (superconvergent patch) recovery procedures which converge faster towards the target errors. The locally smoothed gradients available from this recovery process would be well suited for use in Equation (6.37). Tessler *et. al.* [19] have presented a recovery procedure based on minimisation of a discrete least-squares/penalty constraint functional which produces a smoothed field approaching C^1 continuity for large penalty numbers. This method appears to be better than the superconvergent patch recovery method [18] for linear and cubic elements, however, it requires the solution of a large system of equations as it a global recovery procedure similar to the one described in this chapter.

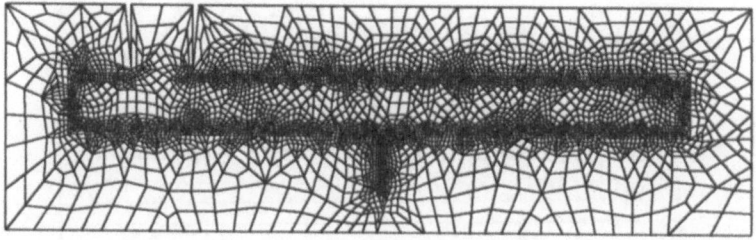

Figure 6.12: Mesh obtained after two adaptive passes on the initial conditions

Returning to the transient casting problem of Figure 6.10, the initial unstructured finite element mesh of linear quadilateral elements is shown in Figure 6.11. Note that the fin area has been refined to ensure the presence of node points in the interior of the fin. The initial conditions were imposed on this mesh and two passes of the error estimator were used on the initial temperature field (without advancing in time) to obtain the mesh shown in Figure 6.12. This procedure ensured that minimal errors were introduced in the problem at the start. The temperature at the mould boundaries was fixed at $20°C$ and the casting boundaries were assumed to be insulated. The adaptive meshes and the results at three different times are shown in Figure 6.13. These results were based on a target error of 25.0%. The mesh refinement can be seen to be concentrated in the areas of high temperature gradient located at the metal-mould interface. A good temperature resolution was achieved at the fin (which is only 1mm wide) as due care was taken to ensure a fine mesh in that region at the beginning. The mesh can be seen to be evolving with the shifting gradients. The solidification range was from $615°C$ to $550°C$. In Figure 6.13(b) almost all of the casting has begun solidifying (below $615°C$). In Figure 6.13(c) most of the casting has solidified (below $550°C$).

Finally, it is important to understand that the errors calculated by the error estimation procedure are obtained from a smoothed solution on the same mesh. Therefore, one may discover lower errors for the same problem for a coarse mesh than for a better fine mesh. In the above problem the errors were higher than 25.0% for all of the results shown as the size of the elements is restricted to a specified minimum in program **HADAPT**. However, we still obtain a very good quality solution without having to refine the whole mesh indiscriminately.

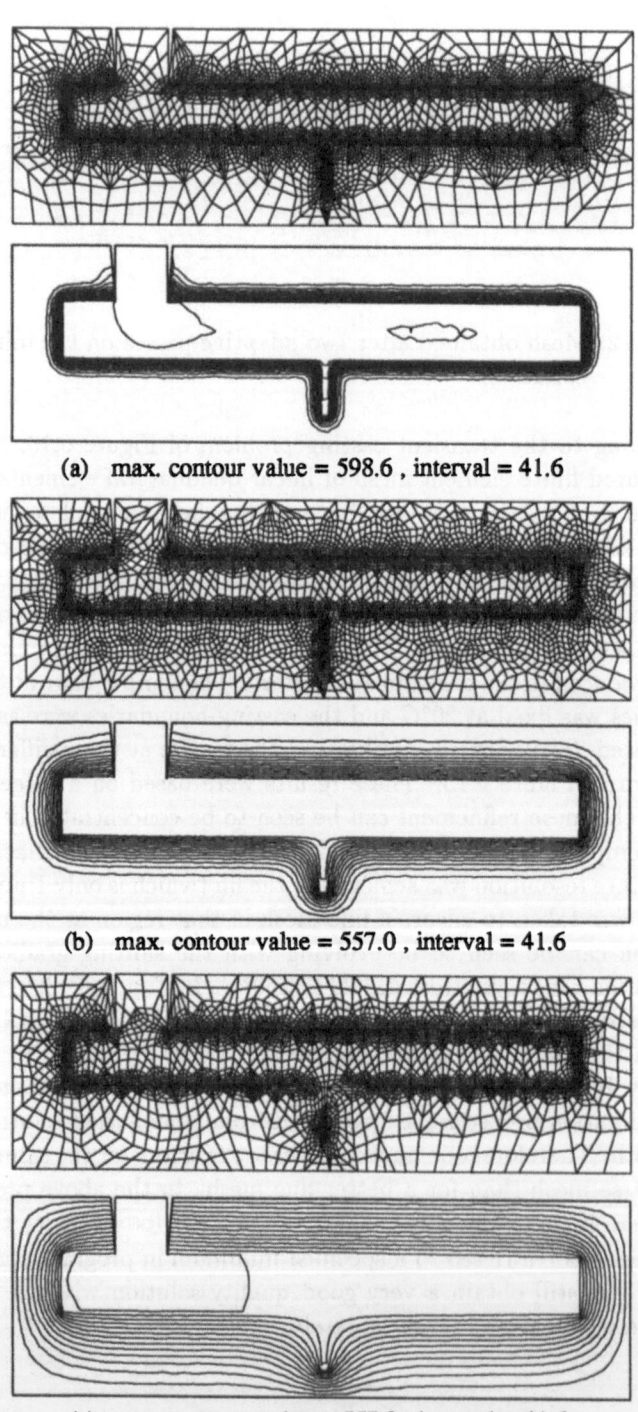

(a)   max. contour value = 598.6 , interval = 41.6

(b)   max. contour value = 557.0 , interval = 41.6

(c)   max. contour value = 557.0 , interval = 41.6

Figure 6.13: Meshes and temperature contours at (a) 3.28 sec,
(b) 16.55 sec, (c) 76.55 sec

References

[1] I. Babuska and W.C. Rheinboldt. Adaptive approaches and reliability estimates in finite element analysis. *Computer Methods in Applied Mechanics and Engineering*, 17/18:519–514, 1979.

[2] I. Babuska and W.C. Rheinboldt. Error estimates for adaptive finite element computations. *SIAM Journal of Numerical Analysis*, 15, 1978.

[3] D.W. Kelly, J.P. De S.R. Gago, O.C. Zienkienwicz, and I. Babuska. A posteriori error analysis and adaptive processes in the finite element method: part 1 — error analysis. *International Journal for Numerical Methods in Engineering*, 19:1596–1619, 1983.

[4] J.P. De S.R. Gago, D.W. Kelly, O.C. Zienkienwicz, and I. Babuska. A posteriori error analysis and adaptive processes in the finite element method: part 2 — adaptive mesh refinement. *International Journal for Numerical Methods in Engineering*, 19:1621–1656, 1983.

[5] J. Peraire, M. Vahdati, K. Morgan, and O.C. Zienkiewicz. Adaptive remeshing for compressible flow computations. *Journal of Computational Physics*, 72:449–466, 1987.

[6] J. Peraire, K. Morgan, and J. Piero. Unstructured finite element mesh generation and adaptive procedures for CFD. In *AGARD FDP: Specialists meeting*, Loen, Norway, May 1989.

[7] O.C.Zienkiewicz, Y.C.Liu, and G.C.Huang. Error estimates and convergence rates for various incompressible elements. *International Journal for Numerical Methods in Engineering*, 28:2191–2202, 1989.

[8] H.C.Huang and R.W.Lewis. Adaptive analysis for heat flow problems using error estimation techniques. In *Sixth International Conference for Numerical Methods in Thermal Problems*, Swansea, U.K., July 1989. Pineridge Press, Swansea.

[9] O.C. Zienkiewicz and J.Z. Zhu. A simple error estimator and adaptive procedure for practical engineering analysis. *International Journal for Numerical Methods in Engineering*, 24:337–357, 1987.

[10] E.Hinton and J.S.Campbell. Local and global smoothing of discontinuous finite element functions using a least squares method. *International Journal for Numerical Methods in Engineering*, 8:461–480, 1974.

[11] T.J.R.Hughes. *The Finite Element Method - Linear Static and Dynamic Finite Element Analysis.* Prentice-Hall International, Inc., Englewood Cliffs, New Jersey 07632, 1987.

[12] M. Zlamal. Superconvergence and reduced integration in the finite element method. *Mathematics of Computation*, 32:663–685, 1978.

[13] O.C.Zienkiewicz and J.Z.Zhu. Adaptivity and mesh generation. *International Journal for Numerical Methods in Engineering*, 32:783–810, 1991.

[14] R.W.Lewis, H.C.Huang, A.S.Usmani, and J.T.Cross. Finite element analysis of heat transfer and flow problems using adaptive remeshing including application to solidification. *International Journal for Numerical Methods in Engineering*, 1991. To be published in a special issue on adaptivity and meshing.

[15] J.Barlow and G.A.O.Davies. Selected FE benchmarks in structural and thermal analysis. Technical Report FEBSTA REV 1, NAFEMS, 1986.

[16] H.S.Carslaw and J.C.Jaeger. *Conduction of Heat in Solids.* Clarendon Press, Oxford, 1959.

[17] P.A.B.de Sampaio, P.R.M.Lyra, K.Morgan, and N.P.Weatherill. Petrov-galerkin solutions of the incompressible Navier-Stokes equations in primitive variables with adaptive remeshing. Technical Report CR/701/92, Department of Civil Engineering, University College of Swansea, 1992.

[18] O.C.Zienkiewicz and J.Z.Zhu. The superconvergent patch recovery and *a posteriori* error estimates. Part 1: The recovery technique, Part 2: Error estimates and adaptivity. *International Journal for Numerical Methods in Engineering*, 33:1331–1382, 1992.

[19] A.Tessler, H.R.Riggs, and S.C.Macy. A variational method for finite element stress recovery and error estimation. *Computer Methods in Applied Mechanics and Engineering*, 111:369–382, 1994.

Chapter 7

Effects of Convection in Heat Transfer

7.1 Introduction

When there is a velocity field present in the problem domain, heat is transported by the medium of convection as well as diffusion. The differential equation governing convective-diffusive heat transfer is reproduced here from Chapter 2.

$$\rho c \frac{DT}{Dt} = \nabla \cdot k \nabla T + Q \tag{7.1}$$

where

$$\frac{DT}{Dt} = \frac{\partial T}{\partial t} + u \frac{\partial T}{\partial x} + v \frac{\partial T}{\partial y} + w \frac{\partial T}{\partial z}$$

As mentioned in Chapter 2, we assume a given velocity field in this text. For the derivations in this chapter we assume a velocity field which is constant with time. This however does limit the applicability of the derivations to a constant velocity field. For fully coupled heat transfer and flow problems, such as natural convection, the Navier-Stokes equations must also be solved to obtain the variable velocity field. For more information on this the reader may refer to the following references [1, 2, 3, 4] and texts such as [5, 6].

The finite element discretisation of the Equation 7.1 via the conventional Galerkin weighted residual method yields a spatially discretised set of equations as for the pure conduction energy equation, with only one additional convective component in the general stiffness matrix for an element e. The *conductivity matrix* for an element e, K_{ij}^e is given in Chapter 3 as,

$$K_{ij}^e = \int_{\Omega_e} \left[\frac{\partial N_i^e}{\partial x} k \frac{\partial N_j^e}{\partial x} + \frac{\partial N_i^e}{\partial y} k \frac{\partial N_j^e}{\partial y} \right] dx dy \tag{7.2}$$

To this we add a *convection matrix* for an element e, $\mathbf{A}_{ij}^e$, as,

$$\mathbf{A}_{ij}^e = \int_{\Omega_e} \rho c \left[N_i^e u \frac{\partial N_j^e}{\partial x} + N_i^e v \frac{\partial N_j^e}{\partial y} \right] dx dy \qquad (7.3)$$

where, u and v are the Cartesian components of the velocity vector. When numerical integration is used to calculate $\mathbf{A}_{ij}^e$, u and v may be calculated at the integration points as $\sum N_k u_k$ and $\sum N_k v_k$, where k is the summation index over the number of nodes in each element. All other terms of the spatially discretised convection-diffusion equation remain the same as derived earlier for the pure conduction case.

The relative amounts of heat transported by the two mechanisms (convection and diffusion), is normally expressed by means of a dimensionless parameter known as the *Peclet number* (Pe). This is defined as,

$$Pe = \frac{\rho c \mid \mathbf{v} \mid L}{k} \qquad (7.4)$$

where L is a characteristic length and $\mid \mathbf{v} \mid$ is the magnitude of velocity. For low values of Pe, the Galerkin finite element discretization of the convection-diffusion equation behaves reasonably well and acceptable solutions are achieved. However, when the convective effect starts to dominate for a given problem *i.e.*, the local values of Pe increase beyond a certain point, spurious oscillations are generated which may completely mask the true solution. Galerkin discretization is therefore no longer sufficient to give reliable or indeed, meaningful results. This is a very active area of research for people involved in computational methods for heat transfer problems or for that matter, the transport of other quantities such as, mass, momentum and pollutants etc. In these problems the term 'convection' which refers to heat transfer is replaced by the more general term 'advection'.

Because of the central difference nature of GFEM approximations, oscillations result in the solution when the finite element mesh is of insufficient refinement to model the variations of the field variable, such as one linear element spanning a region where the field variable varies quadratically. In most cases such spurious oscillations may be eliminated by a judicious mesh refinement in the zones where regions of high gradient are expected. Gresho *et. al.* [7] have shown that in fact GFEM continues to give reliable solutions for Peclet numbers much larger than one (the theoretical limit). They have further shown that the actual limit at which the wiggles appear is related to the product of the solution gradient and the local or mesh Peclet number. The mesh Peclet number is obtained for any element by replacing the characteristic length (L) in Equation (7.4) with the element size (h). Therefore meaningful solutions using GFEM are still obtainable for $Pe \gg 1$ if the mesh is sufficiently refined in the regions of high gradient. This is obviously due to the reduction in the local Pe due to reduced element size. We can therefore conclude that an h-adaptive finite element method combined with GFEM is another way of dealing with advection dominated problems.

In transient problems a timestep size larger than the limits dictated by the physical properties and the spatial step size (mesh size) also results in an oscillatory solution, which may become totally meaningless if the magnitude of the oscillations masks the real values. The remedies of spatial and temporal refinement work well for most diffusion dominated problems, however, when problems of pure advection or advection dominated transport are attempted, very often the conventional remedies of refining the mesh or reducing the timestep size cease to be practical.

In the following sections we describe the two most appropriate (in our view) methods for solving advection dominated problems. The first method is for steady state problems and the other is for transient problems.

7.2   Steady State Advection-diffusion

The basic reason for the failure of GFEM in advection dominated situations is that GFEM is optimal for self-adjoint problems, such as heat conduction which give rise to symmetric stiffness matrices. Advection-diffusion problems are not self-adjoint and produce non-symmetric stiffness matrices. Therefore many methods that have been developed for modelling advection diffusion start by attempting to regain the self adjoint property in the discretized equations. The problem of oscillations in advection dominated flow has been one of the most active areas of research in the numerical modelling of transport phenomena. The period of this research goes back several decades in the case of finite difference modelling and about fifteen years in finite elements. The most conspicuous of all the different techniques proposed to deal with this problem has been the method of upstream differencing, or as more widely referred to, 'upwind methods'. In finite element circles this method was first used by Christie *et. al.* [8] who used modified weighting functions to achieve the upwind effect compared to the conventional Galerkin technique of using the trial functions as the weighting functions. Heinrich *et. al.* [9] generalised the method to two dimensions using a consistent Petrov-Galerkin formulation in which the modified weighting functions were applied to all the terms rather than to the advection term only. The method was further improved by the introduction of 'streamline upwinding' by Brooks and Hughes [10], who removed the problem of excessive diffusion in the crosswind or transverse direction to the flow, by replacing the scalar artificial diffusivity with a diffusivity tensor with non-zero values only in the flow direction. Kelly *et. al.* [11] developed a similar procedure and termed it 'anisotropic balancing dissipation'. There was much criticism of the upwinding methods [7, 12] initially, due to the fact that the effect of upwinding is equivalent to adding artificial diffusion to the standard Galerkin formulation. This, in the early upwind methods manifested itself as excessive crosswind diffusion, when these methods were applied to multi-dimensional problems and problems with transient and source terms. However, in the present stage of development the upwind technique, identified by Hughes [13] by the acronym SUPG (streamline upwind Petrov-Galerkin), has effectively answered most of the criticisms levelled at the early upwind methods. Furthermore, it has proven its effectiveness in providing oscillation free solutions to many advection dominated problems and a sound mathematical basis for this method has also been developed [13]. We describe the implementation of the SUPG method in the following lines.

7.2.1   SUPG Method for Steady Advection-diffusion

We begin with the weighted residual statement for the advection-diffusion equation,

$$\int_{\Omega} W \left[\rho c \mathbf{v} \cdot \boldsymbol{\nabla} \bar{T} - \boldsymbol{\nabla} \cdot k \boldsymbol{\nabla} \bar{T} + Q \right] d\Omega = 0 \tag{7.5}$$

The same is done to the boundary conditions, where the boundary conditions are the same as described in Chapter 2. As in Chapter 3 we require the weighting functions to be zero on the boundary Γ_T where the Dirichlet or fixed boundary condition exists. On the boundary Γ_q we write,

$$\int_{\Gamma_q} \bar{W} \left[k \frac{\partial \bar{T}}{\partial n} + \bar{q} \right] d\Gamma = 0 \tag{7.6}$$

As before the boundaries satisfy,

$$\Gamma_q \cup \Gamma_T = \Gamma$$
$$\Gamma_q \cap \Gamma_T = \Phi$$

where, Φ is the null set.

It is at this stage that we make a break with GFEM and define a new set of Petrov-Galerkin weighting functions,

$$W = N + p \tag{7.7}$$

where N are our usual C^o continuous trial functions and p are perturbation functions which are discontinuous C^{-1} at the boundaries. We now apply the Green's function or integration by parts to the expressions of Equations (7.5) and (7.6). However, since the perturbation functions are discontinuous at element boundaries the integration by parts is only applied to the Galerkin weighted part of the diffusion term. The resulting expression after setting $\bar{W} = -N$ on Γ_q is,

$$\int_\Omega W \left(\rho c \mathbf{v} \cdot \boldsymbol{\nabla} \bar{T} \right) d\Omega + \int_\Omega \boldsymbol{\nabla} N \cdot k \boldsymbol{\nabla} \bar{T} d\Omega \; - \; \sum_{e=1}^M \int_{\Omega_e} p \left[\boldsymbol{\nabla} \cdot k \boldsymbol{\nabla} \bar{T} \right] d\Omega_e$$

$$= \int_{\Gamma_q} N \bar{q} d\Omega - \int_\Omega W Q d\Omega \tag{7.8}$$

where, M is the number of elements. Examining the third term involving the perturbation weighting function p we find that it is applied to the diffusion term of the advection-diffusion differential equation, containing second derivatives. However, as this weighting applies only on an element by element basis, we can make certain simplifications according to Hughes and Brooks [10]. Let us assume the thermal conductivity k to be a constant (as has been done so far in this derivation). If simplex elements are used (linear triangles or linear tetrahedra), then we see that the second derivatives will be zero and therefore this term would vanish. The same is true for bilinear isoparametric elements of rectangular shape and corresponding 3-D elements. For general isoparametric elements with linear interpolation this term may be neglected if the element shapes are not too distorted. Brooks and Hughes [10] report that in higher order elements the contribution of this term may be significant for diffusion dominated cases but for advection dominated cases there is justification for neglecting it.

Having completed a consistent weighted residual formulation using a Petrov-Galerkin type weighting function, we now turn to establishing the nature of the perturbation function p. This function is defined as,

$$p = \frac{\tilde{k}}{\|\mathbf{v}\|} \mathbf{v} \cdot \boldsymbol{\nabla} N \tag{7.9}$$

where $\tilde{k}$ is a scalar controlling the amount of diffusion to be added and $\|\mathbf{v}\|$ is the euclidean norm or the average velocity over the element. This form of the perturbation function has evolved from the early days of upwinding when only a scalar diffusion ($\tilde{k}$) was added to the physical diffusion. In the present form, the function p ensures that $\tilde{k}$ is applied only in the flow or streamline direction to eliminate 'crosswind diffusion'. This form of p results in $\tilde{k}$ being applied as part of a diffusivity (or conductivity, as derived here) tensor. Expanding Equation (7.8) after removing the third term, we have

$$\int_\Omega N \left(\rho c \mathbf{v} \cdot \boldsymbol{\nabla} \bar{T} \right) d\Omega + \int_\Omega \left(\frac{\tilde{k}}{\|\mathbf{v}\|} \mathbf{v} \cdot \boldsymbol{\nabla} N \right) \left(\rho c \mathbf{v} \cdot \boldsymbol{\nabla} \bar{T} \right) d\Omega$$

$$+ \int_\Omega \boldsymbol{\nabla} N \cdot k \boldsymbol{\nabla} \bar{T} d\Omega = \int_{\Gamma_q} N \bar{q} d\Omega - \int_\Omega N Q d\Omega - \int_\Omega \left(\frac{\tilde{k}}{\|\mathbf{v}\|} \mathbf{v} \cdot \boldsymbol{\nabla} N \right) Q d\Omega \tag{7.10}$$

The second term on the LHS and the last term on the RHS of the above equation represent the additional terms obtained from the Petrov-Galerkin formulation. The former represents a stabilising *artificial diffusion* and the latter is the result of a consistent weighting of the thermal load Q. The artificial diffusion term may be written as

$$\int_\Omega \rho c \frac{\tilde{k}}{\|\mathbf{v}\|} \nabla N \cdot \mathbf{v}\mathbf{v}^T \nabla \bar{T} d\Omega \tag{7.11}$$

This expression shows very clearly the structure of the artificial diffusion term as a result of the SUPG formulation. The scalar multipliers combined with the velocity tensor $(\mathbf{v}\mathbf{v}^T)$ produce a kind of conductivity tensor. We now approximate $\bar{T}$ using the trial functions N and convert Equation (7.10) into a finite element algebraic system of equations. In doing so we find that in addition to element matrices and vectors of Chapter 3 and the advection matrix of Section 1 of this chapter, we obtain a stabilising matrix $(\mathbf{S}_{ij}^e)$ for the artificial diffusion term above,

$$\mathbf{S}_{ij}^e = \rho c \frac{\tilde{k}}{\|\mathbf{v}\|} \int_{\Omega_e} \left(u^2 \frac{\partial N_i}{\partial x} \frac{\partial N_j}{\partial x} + v^2 \frac{\partial N_i}{\partial y} \frac{\partial N_j}{\partial y} + uv \frac{\partial N_i}{\partial x} \frac{\partial N_j}{\partial y} + vu \frac{\partial N_i}{\partial y} \frac{\partial N_j}{\partial x} \right) d\Omega_e \tag{7.12}$$

where, u and v may be calculated at the integration points as before. The additional term for the thermal loads may be written for an element e as,

$$\frac{\tilde{k}}{\|\mathbf{v}\|} \int_{\Omega_e} \left(u \frac{\partial N_i}{\partial x} + v \frac{\partial N_i}{\partial y} \right) Q dx dy \tag{7.13}$$

In the above formulation the issue of the scalar artificial diffusion coefficient $(\tilde{k})$ remains outstanding. For 1-D advection problems it is defined as

$$\tilde{k} = \frac{vh}{2} \tilde{\xi} \tag{7.14}$$

where,

$$\tilde{\xi} = \coth \frac{Pe}{2} - \frac{2}{Pe} \tag{7.15}$$

Pe being the mesh or local Peclet number. Such a definition of $\tilde{k}$ produces nodally exact solutions, and is thus optimal. For multi-dimensional problems [10] the above definition is generalised in the absence of a more rigorous alternative. This generalisation however, does not seem to be of major consequence to the solution as it has been pointed out by Brooks and Hughes [10] that the structure of the artificial diffusion term of Equation (7.11) is more important than the actual value of the parameter $\tilde{k}$. The 2-D generalisation of Equations (7.14) and (7.15) for an isoparametric quadrilateral element is given as, follows

$$\tilde{k} = \frac{\tilde{\xi} v_\xi h_\xi + \tilde{\eta} v_\eta h_\eta}{2} \tag{7.16}$$

where,

$$\tilde{\xi} = coth\frac{Pe_\xi}{2} - \frac{2}{Pe_\xi}$$

$$\tilde{\eta} = coth\frac{Pe_\eta}{2} - \frac{2}{Pe_\eta} \qquad (7.17)$$

and

$$Pe_\xi = \frac{v_\xi h_\xi}{k}$$

$$Pe_\eta = \frac{v_\eta h_\eta}{k} \qquad (7.18)$$

In the above equations ξ and η refer to the local or natural coordinate axes of an element.

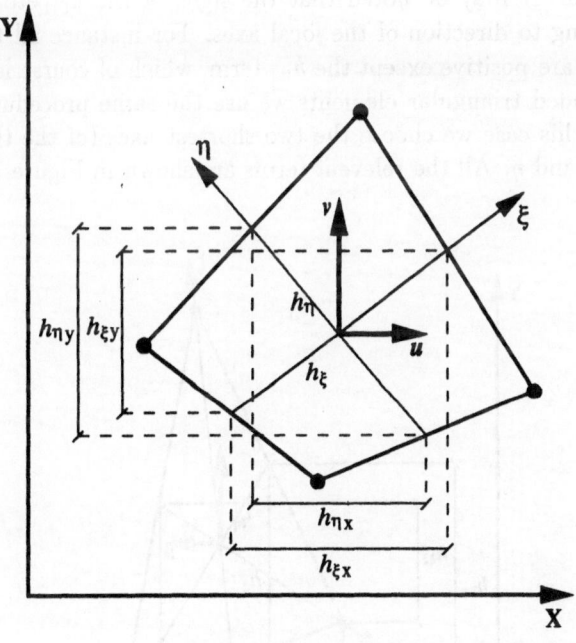

Figure 7.1: Illustration of quantities required to calculate $\tilde{k}$ for the 4-noded isoparametric quadrilateral

An alternative version of the Equations (7.16) to (7.18) which is equivalent and readily programmable may be written for a 4-noded isoparametric quadrilateral element as,

$$\tilde{k} = \frac{\tilde{x}u(h_{\xi x} + h_{\eta x}) + \tilde{y}v(h_{\xi y} + h_{\eta y})}{2} \qquad (7.19)$$

where,

$$\tilde{x} = coth\frac{Pe_x}{2} - \frac{2}{Pe_x}$$
$$\tilde{y} = coth\frac{Pe_y}{2} - \frac{2}{Pe_y} \tag{7.20}$$

and

$$Pe_x = \frac{u(h_{\xi x} + h_{\eta x})}{k}$$
$$Pe_y = \frac{v(h_{\xi y} + h_{\eta y})}{k} \tag{7.21}$$

The definitions of all the above terms are clearly illustrated in Figure 7.1. All the terms in the figure can easily be calculated from the nodal coordinates. It may be noted that the signs of the $h_{\xi x}$ etc. terms will be according to direction of the local axes. For instance all the terms in Figure 7.1 are positive except the $h_{\eta x}$ term, which of course is negative.

For 3-noded triangular elements we use the same procedure as above, except in this case we choose the two shortest axes (of the three natural axes) as ξ and η. All the relevent terms are shown in Figure 7.2.

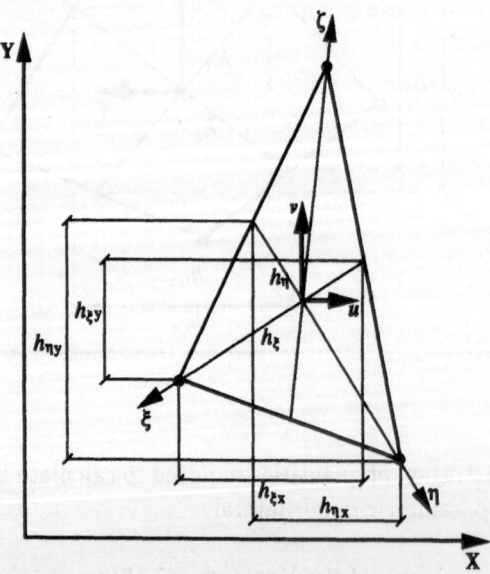

Figure 7.2: Illustration of quantities required to calculate $\tilde{k}$ for the 3-noded triangle

7.2.2 Benchmark Test for the Petrov-Galerkin Method

Figure 7.3 illustrates a typical problem that is widely used to test algorithms for modelling convection dominated algorithms. A uniform flow

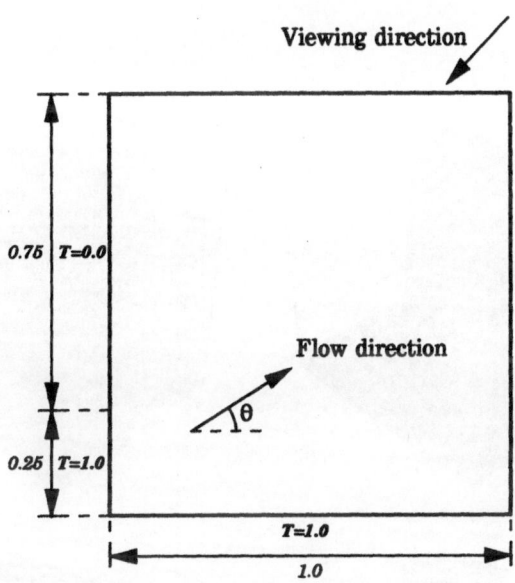

Figure 7.3: Illustration of the benchmark problem to test the Petrov-Galerkin method

field of a unit magnitude is assumed. The inflow boundaries are specified as shown in the figure. The outflow boundaries are considered homogeneous (meaning diffusive flux is zero, which of course means that there is no temperature gradient). The value of θ in Figure 7.3 for the test examples attempted here has been chosen as $30°$. The value of heat capacity (ρc) is set to unity while the value of thermal conductivity (k) is set to 10^{-6}. This low value of conductivity makes this a convection dominated problem with a Pe of 10^6. Figure 7.4 shows the exact solution (E) in comparison to the Galerkin (G), Quadrature Upwind (QU) and SUPG solutions. A uniform mesh of 400 4-noded elements was used for each method. The Quadrature Upwind solution corresponds to the traditional upwind methods which are equivalent to adding a scalar artificial diffusion $\tilde{k}$ regardless of the flow direction. It can be seen clearly from Figure 7.4 that the SUPG method gives us the best results. There are overshoots and undershoots in case of the SUPG method, however these do not affect the solution in other regions. Hughes [13] suggests the use of a discontinuity capturing operator to obtain smoother results.

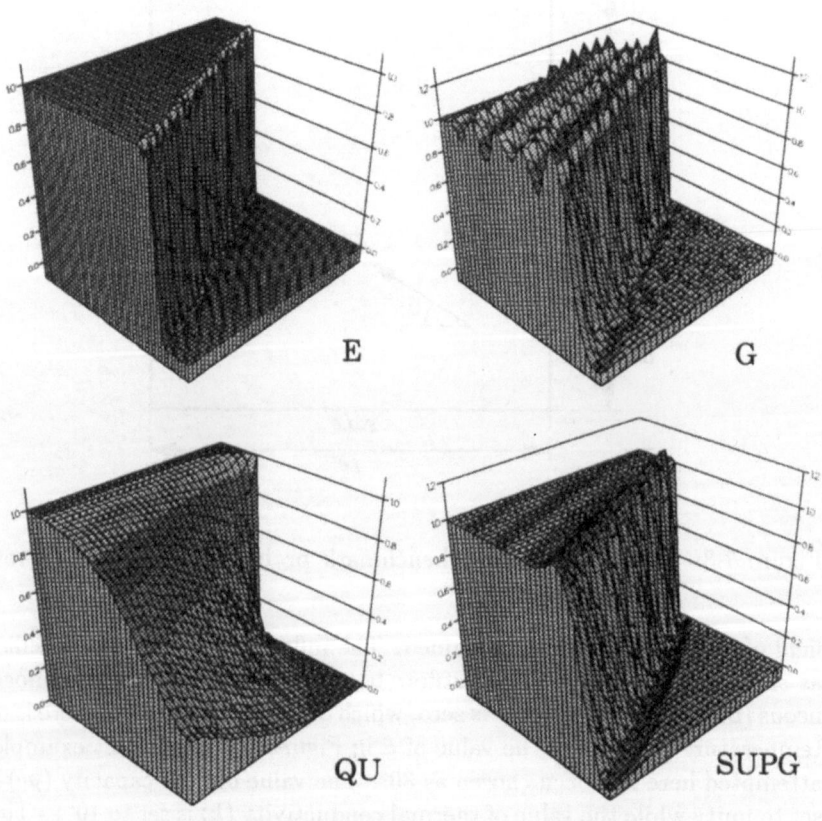

Figure 7.4: Exact solution compared to the Galerkin, Quadrature Upwind
    and SUPG solutions

7.2.3 Adaptive Solution of the Benchmark Problem

The solution to the benchmark problem of the previous section may be further improved by combining the techniques of adaptive mesh refinement discussed in the previous chapter with the SUPG method. Figure 7.5 shows a sequence of meshes which were automatically generated based on a specified target error of 20%. The last mesh at which the target error was achieved clearly shows a band of very fine element marking the region of the highest gradient.

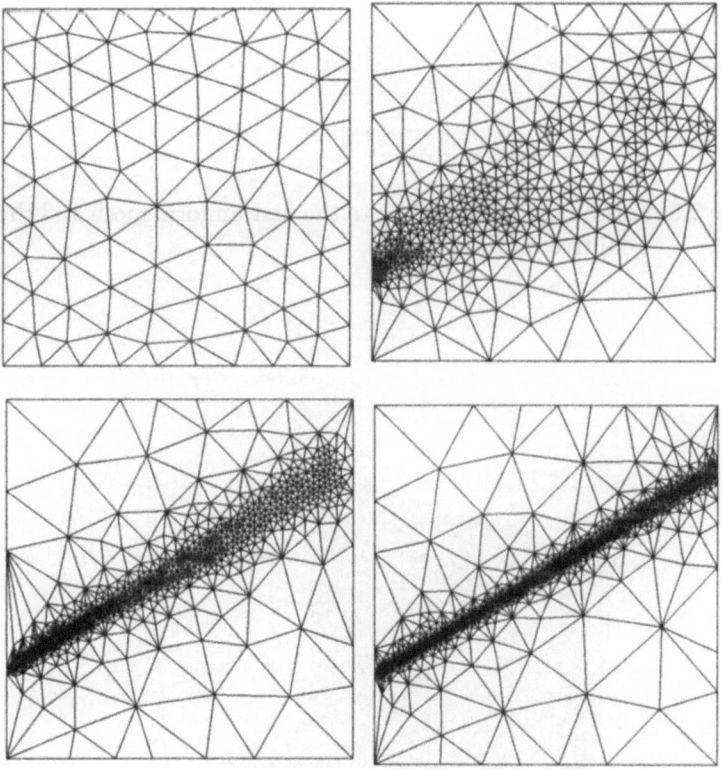

Figure 7.5: The first uniform mesh, two subsequent adaptive meshes and the last mesh at which the target error was achieved

The solutions from the first mesh and the last mesh are shown in Figure 7.6 in the form of line contours. The quality of the adaptive solution can be seen to be far superior.

The converged adaptive solution is again shown in Figure 7.7 for comparison with the solutions in Figure 7.4. It may be observed that the adaptive solution is very close to the exact solution.

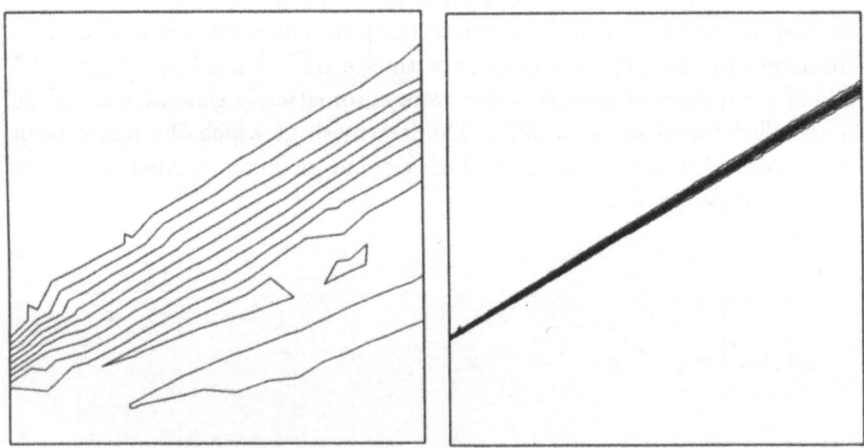

Figure 7.6: Solutions corresponding to the first uniform mesh and the final adaptive mesh

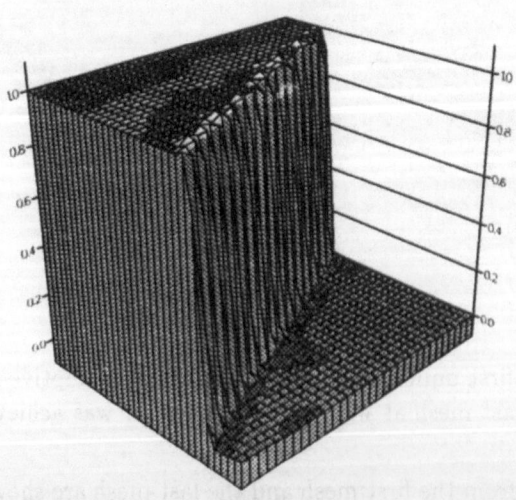

Figure 7.7: 3D illustration of the final adaptive solution

7.3 Transient Advection-diffusion

The SUPG method has essentially been derived based on the steady state advection-diffusion equation and is not strictly applicable to the transient equations [14] being considered here. Furthermore, upwind methods still require empirical factors to control the amount of artificial diffusion as seen in the previous section. However it may still be applied to transient problems with an *ad hoc* definition of the parameter $\tilde{k}$ given by Brooks and Hughes as,

$$\tilde{k} = \frac{\tilde{\xi} v_\xi h_\xi + \tilde{\eta} v_\eta h_\eta}{\sqrt{15}} \tag{7.22}$$

A better alternative to the SUPG method for transient problems, appears to be the Taylor-Galerkin methods introduced by Donea [15] who developed these methods for different time marching schemes. Donea's work was guided by the work of Morton and Parrott [16] who showed that to each particular time stepping method corresponds a different optimal form of Petrov-Galerkin weighting function. This led him to discretise the pure advection equation in time first, with an improved difference approximation of the time derivative term by including higher order Taylor series terms, and then using the conventional GFEM (Bubnov-Galerkin) for spatial discretisation, thus leading to the "Taylor-Galerkin" method. This method is the finite element counterpart of the Lax-Wendroff schemes used in finite differences [17]. Lohner *et. al.* [18] used an Eulerian-Lagrangian approach to arrive at the same system of discretised equations obtained by Donea using the Taylor-Galerkin technique, and justified the use of GFEM for spatial discretisation by writing the advection equation in Lagrangian coordinates and obtaining an adjoint equation for which the conventional GFEM is optimal. This confirmed the characteristic based nature of the Taylor-Galerkin technique. Donea *et. al.* [19] and Zienkiewicz *et. al.* [14, 20] extended their respective methods to solve advection-diffusion problems, using the same basic approach. Gresho *et. al.* [21] derived the same scheme from the argument of 'negative diffusion' that is generated by forward Euler time stepping, and called their correction a 'balancing tensor diffusivity'.

In the subsequent sections the discretised equations will be developed for both the pure advection Equation (7.23) and the advection-diffusion (energy) equation, based on the Taylor-Galerkin method of Donea.

7.3.1   Taylor-Galerkin Method for the Hyperbolic Equation

The first order pure advection equation for the function $F(x, y, t)$, which we call a *pseudo-concentration function*, may be written as,

$$\frac{\partial F}{\partial t} + u\frac{\partial F}{\partial x} + v\frac{\partial F}{\partial y} = 0 \qquad (7.23)$$

This is a hyperbolic equation for which essential boundary conditions may only be applied at inlet portions of the boundary. If such boundary conditions are applied at the outlet a very thin boundary layer appears and extremely fine meshes are required to resolve it. Otherwise, wiggles are generated which are propagated upstream and completely destroy the solution. If conventional GFEM is used to model the advection of the pseudo-concentration function for values of Courant number (C) beyond a certain limit we again encounter the familiar spurious oscillations. Here the Courant number is defined as,

$$C = \frac{v\Delta t}{h} \qquad (7.24)$$

The Taylor-Galerkin method [15] is now applied to Equation (7.23). Writing Equation (7.23) in the form,

$$F' + \nabla \cdot (\mathbf{v}F) = 0 \qquad (7.25)$$

which for incompressible or divergence-free flow $(\nabla \cdot \mathbf{v} = 0)$ becomes

$$F' + (\mathbf{v} \cdot \nabla)F = 0$$

where F' represents the first derivative w.r.t time and so on, and both F and $\mathbf{v}$ are functions of space and time. Using forward time Taylor series, we can write

$$\frac{F_{n+1} - F_n}{\Delta t} = F'_n + \frac{\Delta t}{2}F''_n + O(\Delta t)^2 \qquad (7.26)$$

Donea [15] included a third order term as well, in the above approximation. From Equation (7.25) we have,

$$F'_n = -(\mathbf{v}_n \cdot \nabla)F_n \qquad (7.27)$$

and

$$
\begin{aligned}
F''_n &= -(\mathbf{v}_n \cdot \nabla)F'_n - (\mathbf{v}'_n \cdot \nabla)F_n \\
&= (\mathbf{v}_n \cdot \nabla)(\mathbf{v}_n \cdot \nabla)F_n - (\mathbf{v}'_n \cdot \nabla)F_n \qquad (7.28)
\end{aligned}
$$

Substituting equations (7.27) and (7.28) in Equation (7.26) we obtain the final temporally discretised form as

$$\frac{F_{n+1} - F_n}{\Delta t} = \left(-(\mathbf{v}_n \cdot \nabla) - \frac{\Delta t}{2}(\mathbf{v}'_n \cdot \nabla) + \frac{\Delta t}{2}(\mathbf{v}_n \cdot \nabla)^2\right)F_n \qquad (7.29)$$

which is similar to the expression obtained in [18] using their charateristics based approach. Equation (7.29) can be simplified by approximating $\mathbf{v}'_n$ as $\frac{\mathbf{v}_{n+1}-\mathbf{v}_n}{\Delta t}$, which gives

$$\frac{F_{n+1} - F_n}{\Delta t} = \left(-(\mathbf{v}_{n+\frac{1}{2}} \cdot \nabla) + \frac{\Delta t}{2}(\mathbf{v}_n \cdot \nabla)^2 \right) F_n \qquad (7.30)$$

where $\mathbf{v}_{n+\frac{1}{2}}$ represents $\frac{\mathbf{v}_{n+1}+\mathbf{v}_n}{2}$. Conventional GFEM can now be used to spatially discretise equation (7.30). Using shape functions N_j to approximate F and the same functions N_i as weighting functions, we can write the global system of algebraic equations as,

$$\mathbf{M}\frac{F_{n+1} - F_n}{\Delta t} + (\mathbf{A} + \mathbf{S})F_n = 0 \qquad (7.31)$$

which can be solved for the vector F_{n+1} at each time step as,

$$\left(\frac{\mathbf{M}}{\Delta t}\right)(F_{n+1}) = \left(\frac{\mathbf{M}}{\Delta t} - \mathbf{A} - \mathbf{S}\right)(F_n) \qquad (7.32)$$

where $\mathbf{M}$ may be referred to as a *mass matrix* (it is similar to the *heat capacity matrix* of Chapter 4) and is written as,

$$\mathbf{M} = \sum_{e=1}^{M} \int_{\Omega_e} N_i^e N_j^e dx dy \qquad (7.33)$$

M being the number of elements, $\mathbf{A}$ is

$$\mathbf{A} = \sum_{e=1}^{M} \int_{\Omega_e} -\left(N_i^e(u_{n+\frac{1}{2}})\frac{\partial N_j^e}{\partial x} + N_i^e(v_{n+\frac{1}{2}})\frac{\partial N_j^e}{\partial y} \right) dx dy \qquad (7.34)$$

and $\mathbf{S}$ is a balancing diffusion type part which after integrating the second derivative terms by parts is obtained as

$$\mathbf{S} = \sum_{e=1}^{M} \frac{\Delta t}{2} \int_{\Omega_e} \left(u^2 \frac{\partial N_i^e}{\partial x}\frac{\partial N_j^e}{\partial x} + v^2\frac{\partial N_i^e}{\partial y}\frac{\partial N_j^e}{\partial y} + uv\frac{\partial N_i^e}{\partial x}\frac{\partial N_j^e}{\partial y} + vu\frac{\partial N_i^e}{\partial y}\frac{\partial N_j^e}{\partial x} \right) d\Omega_e$$
$$(7.35)$$

here all the velocities are at time level n. It can be seen from the above derivation that an 'artificial diffusion' or 'balancing diffusion' type term is automatically produced, which has tensorial structure resulting in its influence being limited to the streamline directions. In fact this term is actually the same as the stabilising matrix given by Equation (7.12) derived earlier for the Petrov-Galerkin formulation. The only difference is in the scalar multipliers, which for the Equation (7.12) is obtained in an *ad hoc* manner while here it arises naturally. Donea [15] mentions that this term may not be thought of as artificial diffusion but as part of a more accurate temporal discretisation.

Equation (7.32) is explicit and therefore conditionally stable. It may be solved using a lumped mass matrix, with the advantage of uncoupled equations. A simple way of obtaining a lumped mass matrix is by replacing all diagonal entries of matrix M by the sum of all the terms in the corresponding row. All off-diagonal terms are zero. For the 8-noded quadrilateral element and the 6-noded triangle element such lumping produces negative and zero diagonal entries. Special shape functions are used to remedy this problem [22]. For the 6-noded triangle these shape functions are given in the program accompanying this text. The use of lumped mass matrices for transient advection problems generally degrades the solution by introducing oscillations and phase errors [7]. Donea *et. al.* [19] suggests an iterative explicit procedure which retains the beneficial effects of the consistent mass matrix. This procedure may be applied to an equation system given by,

$$\mathbf{M}\mathbf{u} = \mathbf{f} \tag{7.36}$$

where,

$$\mathbf{u} = \mathbf{u}_{n+1} - \mathbf{u}_n$$

(n being the time level) according to the following relation,

$$\mathbf{L}\mathbf{u}^{p+1} = \mathbf{f} - (\mathbf{M} - \mathbf{L})\mathbf{u}^p \tag{7.37}$$

here, $\mathbf{M}$ and $\mathbf{L}$ are the consistent and lumped mass matrices respectively and p is the iteration index.

As Equation (7.32) only conditionally stable, we can only use time steps below a certain limit. The stability limit for a lumped mass solution is

$$C = \frac{v\Delta t}{h} \leq 1.0 \tag{7.38}$$

and for consistent mass solution is

$$C \leq \frac{1}{\sqrt{3}} \tag{7.39}$$

C being the Courant number, while v is the magnitude of velocity and h is the element size for linear elements. This limit can be very severe specially when consistent mass matrices are used and matrix inversion is necessary for solution.

7.3.2 Benchmark Test for the Taylor-Galerkin Method

The rotating cone problem shown in Figure 7.8 has been widely used to test advection algorithms, and has been recommended strongly for such use, for instance by Gresho and Lee [7]. Figure 7.9 shows the initial configuration, where the cone represents the values of some function which

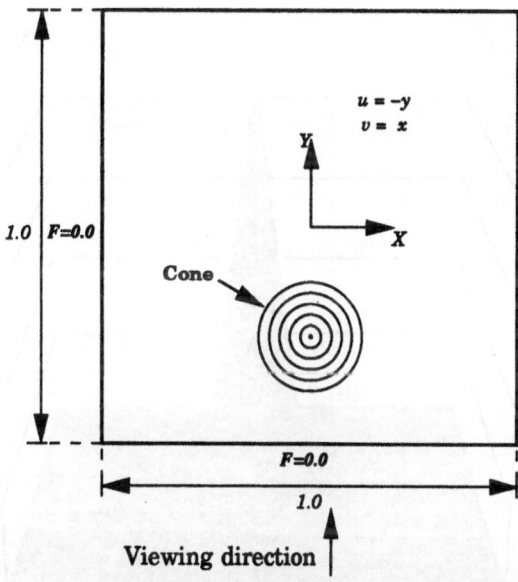

Figure 7.8: Illustration of the benchmark problem to test the Taylor-Galerkin method

must be advected in the square domain (of unit dimensions) by a velocity field of magnitude equal to distance from the centre of the square and of an orbital direction with respect to the centre. This in physical terms means that the cone should be transported unchanged in a circular path about the centre. Figure 7.10 shows the results in the form of cone positions after it has been transported a quarter of a circle. A total of 200 timesteps (for a total time of 2π units) were used to complete the revolution, on a uniform mesh of 1024 4-noded elements. The timestep and mesh size correspond to a Courant number of unity, approximately. The results show a very high quality solution as achieved in [15] and [18], with negligible dissipation and phase errors. Dissipation errors manifest themselves as reduction in cone height. In the example shown the cone retains more than 90% of its height. Phase errors are manifested in a lag of the advected field. There is no discernible lag in the results of Figure 7.10. Slight wiggles are present at the leading and trailing edges of the cone, however these are essentially local without any significant effect in the rest of the domain, as we observed for the SUPG method in the steady state example. Such wiggles are expected in higher order solution of linear problems. An iterative explicit method was used for solving the equation system in this problem as explained in the previous section.

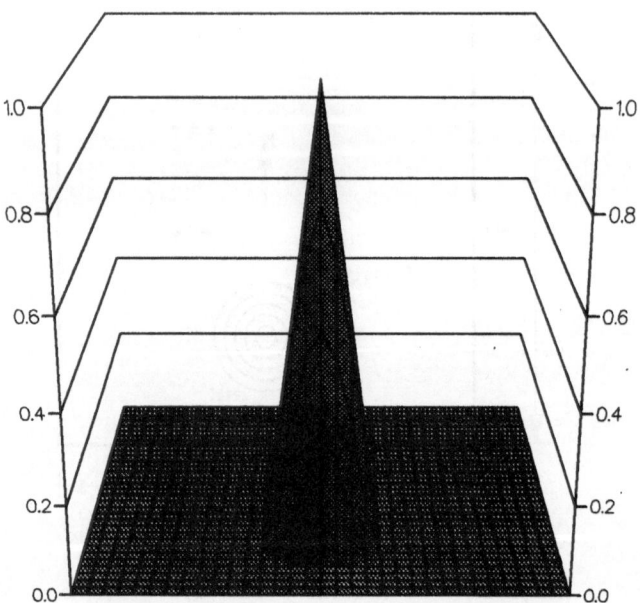

Figure 7.9: Initial configuration for the rotating cone problem.

7.3.3 Taylor-Galerkin Method for Coupled Advection-diffusion

In this section the Taylor-Galerkin method is applied to the energy equation based on references [19, 14, 20]. Writing the energy conservation equation with the source term Q (which is the rate of internal heat generation)

$$\rho c\,(T' + \mathbf{v} \cdot \boldsymbol{\nabla} T) = \boldsymbol{\nabla} \cdot k\,\boldsymbol{\nabla} T + Q \qquad (7.40)$$

where T' represents the first derivative w.r.t time and so on. Using forward time Taylor series as before, we write

$$\frac{T_{n+1} - T_n}{\Delta t} = T'_n + \frac{\Delta t}{2} T''_n + O(\Delta t)^2 \qquad (7.41)$$

From Equation (7.40) we obtain

$$\rho c T'_n = -\rho c(\mathbf{v}_n \cdot \boldsymbol{\nabla})T_n + \boldsymbol{\nabla} \cdot k\,\boldsymbol{\nabla} T_n + Q_n \qquad (7.42)$$

and

$$\rho c T''_n = -\rho c\,((\mathbf{v}_n \cdot \boldsymbol{\nabla})T'_n + (\mathbf{v}'_n \cdot \boldsymbol{\nabla})T_n) + \boldsymbol{\nabla} \cdot k\,\boldsymbol{\nabla} T'_n + \dot{Q}_n \qquad (7.43)$$

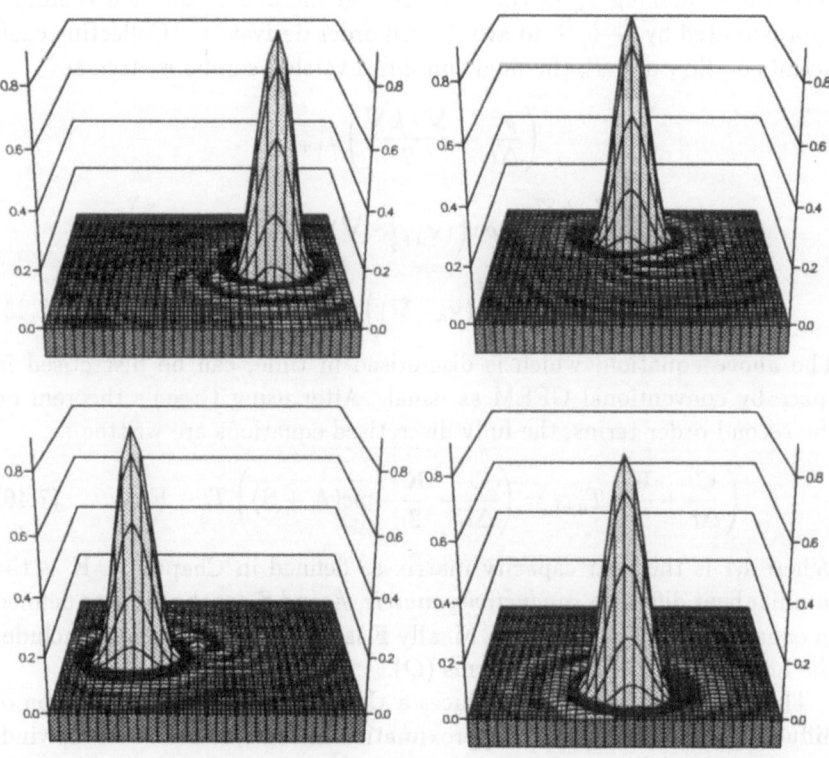

Figure 7.10: Explicit solution for the rotating cone problem showing the cone from after a quarter revolution, top-left, to a full revolution, bottom right.

Substituting equations (7.42) and (7.43) in Equation (7.41) after approximating $\mathbf{v}_n'$ as $\frac{\mathbf{v}_{n+1}-\mathbf{v}_n}{\Delta t}$, we obtain

$$\rho c\frac{T_{n+1}-T_n}{\Delta t} = -\rho c(\mathbf{v}_{n+\frac{1}{2}} \cdot \boldsymbol{\nabla})T_n + \boldsymbol{\nabla}\cdot k\,\boldsymbol{\nabla}T_n + Q_n$$
$$+ \frac{\Delta t}{2}\left(-\rho c(\mathbf{v}_n\cdot\boldsymbol{\nabla})T_n' + \boldsymbol{\nabla}\cdot k\,\boldsymbol{\nabla}T_n' + \dot{Q}_n\right) \tag{7.44}$$

The T_n' in the fourth term can be approximated by Equation (7.42), however, the remaining T_n' in the fifth term in the above equation is simply approximated by $\frac{T_{n+1}-T_n}{\Delta t}$ to avoid third order derivatives. Collecting coefficients of T_{n+1} and T_n the final simplified version can be written as

$$\left(\frac{\rho c}{\Delta t} - \frac{\boldsymbol{\nabla}\cdot k\,\boldsymbol{\nabla}}{2}\right)T_{n+1}$$

$$= \left(\frac{\rho c}{\Delta t} + \frac{\boldsymbol{\nabla}\cdot k\,\boldsymbol{\nabla}}{2} + \rho c\left((\mathbf{v}_{n+\frac{1}{2}}\cdot\boldsymbol{\nabla}) + \frac{\Delta t}{2}(\mathbf{v}_n\cdot\boldsymbol{\nabla})^2\right)\right)T_n$$

$$+ \left(1 - \frac{\Delta t}{2}(\mathbf{v}_n\cdot\boldsymbol{\nabla})\right)Q_n + \frac{\Delta t}{2}\dot{Q}_n \tag{7.45}$$

The above equation, which is discretised in time, can be discretised in space by conventional GFEM as usual. After using Green's theorem on the second order terms, the fully discretised equations are written as,

$$\left(\frac{\mathbf{C}}{\Delta t} + \frac{\mathbf{K}}{2}\right)T_{n+1} = \left(\frac{\mathbf{C}}{\Delta t} - \frac{\mathbf{K}}{2} - \rho c(\mathbf{A}+\mathbf{S})\right)T_n + \mathbf{F} \tag{7.46}$$

Where $\mathbf{M}$ is the heat capacity matrix as defined in Chapter 4, $\mathbf{K}$ is the familiar heat diffusion *conductivity matrix*, $\mathbf{A}$ and $\mathbf{S}$ are the same as defined in equations (7.34) and (7.35). Finally $\mathbf{F}$ is the load vector which includes the internal heat generation terms (Q).

The above discretisation produces a Crank-Nicholson approximation of diffusion while the advection approximation includes as before an 'upwind' or 'balancing diffusion' type term. The expression obtained here, corresponds exactly to Zienkiewicz *et. al.* [14, 20], who used a characteristics based approach in their derivation.

In spite of the Crank-Nicholson type approximation of diffusion, Equation (7.46) remains explicit for advection and is thus conditionally stable. The stability limit for the above expression is the same as for Equation (7.32) as the thermal diffusion part above is unconditionally stable. If the parameter α from the generalized midpoint rule of Chapter 4 is introduced to the diffusion part as in [14, 20], we may obtain a fully explicit system by setting α equal to zero, resulting in,

$$\frac{\mathbf{C}}{\Delta t}T_{n+1} = \left(\frac{\mathbf{C}}{\Delta t} - \mathbf{K} - \rho c(\mathbf{A}+\mathbf{S})\right)T_n + \mathbf{F} \tag{7.47}$$

The stability limits for various values of α from a linear stability analysis are reported in [14, 20] as,

$$C \leq -\frac{1-2\alpha}{Pe} + \sqrt{\frac{1}{3} + \left(\frac{1-2\alpha}{Pe}\right)^2} \tag{7.48}$$

for consistent capacity matrices, and

$$C \leq -\frac{1-2\alpha}{Pe} + \sqrt{1 + \left(\frac{1-2\alpha}{Pe}\right)^2} \tag{7.49}$$

for lumped capacity matrices. Here C is the Courant number and Pe is the Peclet number

For most high Pe problems it is economical to use the explicit version of the discrete system given by Equation 7.47. To take full advantage of the explicit system the use of lumped mass or capacity matrices is almost mandatory. This however, as mentioned earlier introduces oscillations and phase errors in the solution of advection problems. For pure diffusion the reverse is true, *i.e.* mass lumping actually produces smoother results. In light of this fact Donea [15] proposed the separation of the two parts to achieve a much improved solution. This in context of Equation 7.47 may be accomplished in the following two stages:

Convection

$$\frac{\mathbf{C}}{\Delta t}T^* = \left(\frac{\mathbf{C}}{\Delta t} - \rho c(\mathbf{A} + \mathbf{S})\right) T_n \tag{7.50}$$

Here T^* is an intermediate temperature field obtained by an iterative explicit solution of only the advection part of Equation 7.47.

Diffusion

$$\frac{\mathbf{C}_L}{\Delta t}T_{n+1} = \left(\frac{\mathbf{C}}{\Delta t} - \mathbf{K}\right) T^* + \mathbf{F} \tag{7.51}$$

Solving the diffusion part as above gives the final temperature field. Here a straightforward explicit solution using a lumped capacity matrix ($\mathbf{C}_L$) gives a smooth solution.

Advective boundary conditions for the above problem are applied in the convection stage and diffusive conditions are applied in the diffusion stage. The results obtained using the two-stage procedure above are much smoother than if Equation 7.47 is used directly. As with all splitting methods some splitting errors are introduced here, however, the benefits seem to outweigh the costs significantly.

References

[1] P.M.Gresho, R.L.Lee, and R.L.Sani. On the time-dependent solution of the incompressible Navier-Stokes equations in two and three dimensions. In *Recent Advances in Numerical Methods in Fluids*, volume 1. Pineridge Press Limited, Swansea, 1980.

[2] P.M.Gresho, R.L.Lee, S.T.Chan, and R.L.Sani. Solution of the time-dependent Navier-Stokes and Boussinesq equations using the Galerkin finite element method. In *Proceedings of the IUTAM Symposium on Approximation Methods for Navier-Stokes Problems*, Paderborn, West Germany, September 1979. Springer-Verlag.

[3] P.M.Gresho. On the theory of semi-implicit projection methods for viscous incompressible flow and its implementation via a finite element method that also introduces a consistent mass matrix. part 1: Theory and part 2: Implementation. *International Journal for Numerical Methods in Fluids*, 11:587–659, 1990.

[4] J.Donea, S.Giuliani, H.Laval, and L.Quartapelle. Finite element solution of the unsteady Navier-Stokes equations by a fractional step method. *Computer Methods in Applied Mechanics and Engineering*, 30:53–73, 1982.

[5] O.C.Zienkiewicz and R.L.Taylor. *The Finite Element Method: Volumes 1 and 2*. McGraw-Hill Book Company, London, 1987.

[6] C.Taylor and T.G.Hughes. *Finite Element Programming of the Navier-Stokes Equations*. Pineridge Press, Swansea, U.K., 1981.

[7] P.M. Gresho and R.L.Lee. Don't suppress the wiggles - they are telling you something! *Computers and Fluids*, 9:223–253, 1981.

[8] I.Christie, D.F.Griffiths, A.R.Mitchell, and O.C.Zienkiewicz. Finite element methods for second order differential equations with significant first derivatives. *International Journal for Numerical Methods in Engineering*, 10:1389–1396, 1976.

[9] J.C.Heinrich and O.C.Zienkiewicz. Quadratic finite element schemes for two-dimensional convective transport problems. *International Journal for Numerical Methods in Engineering*, 11:1831–1844, 1977.

[10] A.N.Brooks and T.J.R.Hughes. Streamline upwind/Petrov-Galerkin formulations for convection dominated flows with particular emphasis on the incompressible Navier-Stokes equations. *Computer Methods in Applied Mechanics and Engineering*, 32:199–259, 1982.

[11] D.W.Kelly, S.Nakazawa, O.C.Zienkiewicz, and J.C.Heinrich. A note on anisotropic balancing dissipation in finite element method approximation to convective diffusion problems. *International Journal for Numerical Methods in Engineering*, 15:1705–1711, 1980.

[12] B.P.Leonard. A survey of finite differences of opinion on numerical muddling of the incomprehensible defective confusion equation. In T.J.R.Hughes, editor, *Finite Element Methods for Convection Dominated Flows*, volume 34. ASME, AMD, 1979.

[13] T.J.R.Hughes. Recent progress in the development and understanding of SUPG methods with special reference to the compressible Euler and Navier-Stokes equations. In R.H.Gallagher, R.Glowinski, P.M.Gresho, J.T.Oden, and O.C.Zienkiewicz, editors, *Finite Elements in Fluids*, volume 7. John Wiley and Sons, 1988.

[14] O.C.Zienkiewicz, R.Lohner, K.Morgan, and S.Nakazawa. Finite elements in fluid mechanics - a decade of progress. In R.H.Gallagher, J.T.Oden, O.C.Zienkiewicz, T.Kawai, and M.Kawahara, editors, *Finite Elements in Fluids*, volume 5. John Wiley and Sons, 1984.

[15] J.Donea. A Taylor-Galerkin method for convective transport problems. *International Journal for Numerical Methods in Engineering*, 20:101–119, 1984.

[16] K.W.Morton and A.K.Parrott. Generalized Galerkin methods for first order hyperbolic equations. *Journal of Computational Physics*, 36:249–270, 1980.

[17] P.J.Roache. *Computational Fluid Mechanics*. Hermosa Publishers, Albuquerque, U.S.A., 1976.

[18] R.Lohner, K.Morgan, and O.C.Zienkiewicz. The solution of non-linear hyperbolic equation systems by the finite element method. *International Journal for Numerical Methods in Fluids*, 4:1043–1063, 1984.

[19] J.Donea, S.Giuliani, H.Laval, and L.Quartapelle. Time-accurate solution of advection-diffusion problems by finite elements. *Computer Methods in Applied Mechanics and Engineering*, 45:123–145, 1984.

[20] O.C.Zienkiewicz, R.Lohner, K.Morgan, and J.Peraire. High-speed compressible flow and other advection dominated problems of fluid dynamics. In R.H.Gallagher, G.Carey, J.T.Oden, and O.C.Zienkiewicz, editors, *Finite Elements in Fluids*, volume 6. John Wiley and Sons, 1985.

[21] P.M. Gresho, S.T.Chan, R.L.Lee, and C.D.Upson. A modified finite element method for solving the time-dependent, incompressible Navier-Stokes equations. *International Journal for Numerical Methods in Fluids*, 4:557–598, 1984.

[22] J.Donea, S.Giuliani, and H.Laval. Accurate explicit finite element schemes for convective-conductive heat transfer problems. In T.J.R.Hughes, editor, *Finite Element Methods for Convection Dominated Flows*, volume 34. ASME, AMD, 1979.

Appendix A

Software Description for HEAT2D

A.1 Introduction

In this appendix we present a detailed description of the program **HEAT2D** with complete instructions for the user. This program allows a user to perform steady or transient heat transfer analysis for 2-D plane or axisymmetric problems. The program permits the use of non-linear material properties. Boundary conditions permitted are, fixed temperature, fixed flux, convective flux and radiation to ambient space. The phase change phenomenon may be modelled for the cases of both latent heat release at a fixed temperature and over a range of temperatures. Internal heat generation may be modelled by specifying as thermal loads. Forced convection heat transfer may be modelled by specifying a velocity field as input. The program uses four types of elements, *i.e.*, 3 and 4-noded linear and 6 and 9-noded quadratic elements.

A.2 Glossary of Variable Names

A brief description of the main integer and real variables, main integer and real arrays and the main subroutines is listed in the following sections in alphabetical order.

A.2.1 Main Variables

Variable name	Description
alpha	Time-stepping scheme selector for transient problems (0 to 1)
dtime	Time step size for transient problems
dtmax	Maximum time step size for non-linear transient problems

Variable name	Description
factr	A factor for by which the timestep size is modified
iaxsy	0 for plane and 1 for axisymmetric problems
iconv	0 for conduction only and 1 for conduction and convection
ilinr	0 for linear and 1 for non-linear problems
ipetr	0 for GFEM convection and 1 for SUPG convection
iphas	Material number of the phase change material (0 if none) If enthalpy method is used for phase change set *iphas*=0
itran	0 for steady and 1 for transient problems
mboun	Maximum number of of boundary conditions
mcdpt	Maximum number of points defining property and load variations
mdime	Maximum number of dimensions (2)
melem	Maximum number of elements
mfron	Maximum front width when frontal solver is used (for convection)
mgaus	Maximum number of integration points (7)
mmatr	Maximum number of materials
mnpel	Maximum number of nodes per element (6)
mpoin	Maximum number of nodes
mprof	Maximum profile size
ncdpt	Number of points defining property variations
nelem	Number of elements
neumn	Number of Neumann type boundary conditions
nfixb	Number of fixed temperature nodes
ngaus	Number of integration points
niter	Maximum number of iterations for non-linear problems
nitdn	Iteration number at or above which timestep size is increased
nitup	Iteration number at or below which timestep size is decreased
nmatr	Number of materials
nnpel	Number of nodes per element
nnpfc	Number of nodes per face
npoin	Number of nodes
nprof	Profile size
ntiml	Number of points defining load variations with time
nther	Number of nodes with thermal loads
qlatn	Latent heat per unit volume
relax	Relaxation factor for non-linear problems
stime	Start time for transient problems

Variable name	Description
tliqs	Liquidus temperature
toler	Convergence tolerance for non-linear problems only
tsols	Solidus temperature
ttime	End time for transient problems

A.2.2 Main Arrays

Array name	Description
amass(mnpel,mnpel)	Element capacity matrix
ambit(mboun)	Ambient temperature at the face corresponding to *nfacb*
arwet(mgaus,melem)	Weighting factors at each integration point in an element
astif(mnpel,mnpel)	Element stiffness matrix
capcy(mmatr)	Base heat capacity value for material
cdvlu(mcdpt,mmatr)	Conductivity variation with temperature
coeff(mboun)	Convective heat transfer coefficient at the face corresponding to *nfacb*
condy(mmatr)	Base conductivity value for material
coord(mpoin,mdime)	Nodal coordinates
cpvlu(mcdpt,mmatr)	Capacity variation with temperature
derv1(mnpel,mgaus,melem)	Cartesian derivatives of shape functions with respect to the X-axis
derv2(mnpel,mgaus,melem)	Cartesian derivatives of shape functions with respect to the y-axis
fixed(mboun)	Temperature corresponding to nodes in *iffix*
fluxe(mboun)	Heat flux into the face corresponding to *nfacb*
force(mpoin)	Global right hand side vector
gflum(mfron,mfron)	Global non-symmetric coefficient matrix for convection
gstif(mprof)	Global coefficient matrix in vector form
iffix(mpoin)	Index indicating free/fixed temperature nodes
isoli(mpoin)	Index indicating phase of each node (solid/liquid/mush)
lsoli(mpoin)	Index indicating solidification or melting
lnods(melem,mnpel)	An extra element nodal connectivity array for the front solver
lhedv(mfron)	Array used in the front solver

Array name	Description
locel(mnpel)	Array used in the front solver
mtype(melem)	Element material type
nadfm(mpoin)	First global degree of freedom number at each node
nconc(melem,mnpel)	Element nodal connectivities
ndest(mnpel)	Array used in the front solver
ndiag(mpoin)	Index indicating the diagonal positions in 'gstif'
nelmb(mboun)	Element number with a Neumann type boundary condition
nfacb(mboun)	Face number of the element above
nfixd(mboun)	Node number of fixed temperature boundary
nload(mboun)	Node number of specified thermal load
nodfm(mpoin)	Total number of degrees of freedom at each node (1)
petrv(melem)	Parameter $\tilde{k}$ for the amount of diffusion added in the SUPG method
pnorm(mfron)	Array used in the front solver
qcumu(mpoin)	Cumulative amount of latent heat released or absorbed
qlath(mpoin)	Amount of latent heat processed at each iteration
qtotl(mpoin)	Amount of latent heat available at each node
qincr(mpoin)	Increment to *qlath* at the end of each iteration
qresi(mpoin)	Residual latent heat after phase change has taken place
radia(mboun)	Radiative coefficient at the face corresponding to *nfacb*
shapf(mnpel,mgaus,melem)	Shape functions at each integration point and element
tempr(mpoin)	Current temperature field
tfixd(mpoin)	Temperature corresponding to the fixed nodes
tilod(mcdpt)	Time corresponding to load variation *valod*
tlast(mpoin)	Temperature field at the last time step or iteration
tload(mboun)	Thermal load corresponding to nodes *nload*

Array name	Description
tvalu(mcdpt,mmatr)	Temperature values corresponding to *cpvlu* and *cdvlu* variations
uloca(mnpel)	X-component of the velocity at the nodes of an element
vloca(mnpel)	Y-component of the velocity at the nodes of an element
uvelo(mpoin)	X-component of the velocity specified at nodal points of the mesh
vvelo(mpoin)	Y-component of the velocity specified at nodal points of the mesh
valod(mcdpt)	Thermal load variation with time
volum(mpoin)	Volum of phase change material at each node

A.2.3 Main Subroutines

Subroutine	Description
BOUNDS	Calculates neumann boundary condition effects
FRONTS	Frontal solver for unsymmetric system of equations [1]
JACOBN	Evaluates the jacobian matrix and its determinant
LATENT	Deals with latent heat
LATVEC	Constructs the global nodal latent heat vector
NONLIN	Evaluates nonlinear quantities
OUTPUT	Outputs temperature
PETROV	Calculates the scalar coefficient $\tilde{k}$ for the SUPG method
PHASCH	Calculates the latent heat source term for each iteration
PREWRK	Prepares necessary arrays
PROFAC	Factorizes global stiffness matrix [2] (symmetric)
PROSOL	Profile solver for symmetric system of equations [2]
READIN	Reads all problem data
SHAPEF	Calculates element shape functions and derivatives
SHAPE3	3-noded element shape functions and derivatives
SHAPE4	4-noded element shape functions and derivatives
SHAPE6	6-noded element shape functions and derivatives
SHAPE9	9-noded element shape functions and derivatives
STEADY	Performs steady state analysis
STIFFN	Constructs element stiffness matrix
TEMCOR	Corrects nodal temperatures for consistency with latent heat evolved
TRANSI	Performs transient analysis

A.3   Program Overview

The main routine is listed in the following lines, the program structure
with all the subroutines is shown in Figure A.1.

```
      PROGRAM HEAT2D                                              HEAT   1
C***************************************************************** HEAT   2
C    HEAT TRANSFER ANALYSIS PROGRAM FOR PLANAR AND                HEAT   3
C    AXISYMMETRIC PROBLEMS USING 3 AND 6 NODED TRIANGULAR,        HEAT   4
C    AND 4 AND 9 NODED QUADRILATERAL ELEMENTS.                    HEAT   5
C                                                                 HEAT   6
C                                    A.S USMANI & H.C. HUANG       HEAT   7
C***************************************************************** HEAT   8
      PARAMETER (MELEM=400,MPOIN=441,MNPEL=9,MMATR=1,MBOUN=41,MCDPT=10, HEAT   9
     -           MDIME=2,MGAUS=9,MPROF=9999,MFRON=99)             HEAT  10
      IMPLICIT REAL*8(A-H,O-Z)                                    HEAT  11
      DIMENSION NTYPE(MELEM),NCONC(MELEM,MNPEL),COORD(MPOIN,MDIME), HEAT  12
     -          CONDY(MMATR),CAPCY(MMATR),FIXED(MBOUN),NFIXD(MBOUN), HEAT  13
     -          NFACB(MBOUN),NELMB(MBOUN),NLOAD(MBOUN),FLUXE(MBOUN), HEAT  14
     -          COEFF(MBOUN),RADIA(MBOUN),AMBIT(MBOUN),TLOAD(MBOUN), HEAT  15
     -          TEMPR(MPOIN),CDVLU(MCDPT,MMATR),CPVLU(MCDPT,MMATR), HEAT  16
     -          TVALU(MCDPT,MMATR),TILOD(MCDPT),VALOD(MCDPT)        HEAT  17
      DIMENSION SHAPF(MNPEL,MGAUS,MELEM),DERV1(MNPEL,MGAUS,MELEM),  HEAT  18
     -          DERV2(MNPEL,MGAUS,MELEM),ARWET(MGAUS,MELEM),        HEAT  19
     -          WEIGB(MGAUS),POSGB(MGAUS),WEIGP(MGAUS),POSGX(MGAUS), HEAT  20
     -          POSGY(MGAUS)                                        HEAT  21
      DIMENSION NDFEL(MNPEL),ASTIF(MNPEL,MNPEL),AMASS(MNPEL,MNPEL), HEAT  22
     -          NCOLM(MPOIN),IFFIX(MPOIN),TFIXD(MPOIN),TEMPT(MNPEL), HEAT  23
     -          TEMP1(MNPEL,MNPEL),RVECT(MNPEL),NBELM(MELEM,4),      HEAT  24
     -          XCORD(MNPEL),YCORD(MNPEL),EFORC(MNPEL)              HEAT  25
      DIMENSION ISOLI(MPOIN),LSOLI(MPOIN),ILATN(MPOIN),VOLUM(MPOIN), HEAT  26
     -          QTOTL(MPOIN),QCUMU(MPOIN),QLATH(MPOIN),QINCR(MPOIN), HEAT  27
     -          QRESI(MPOIN)                                        HEAT  28
      DIMENSION GSTIF(MPROF),NDIAG(MPOIN),FORCE(MPOIN),TLAST(MPOIN) HEAT  29
      DIMENSION UVELO(MPOIN),VVELO(MPOIN),PETRV(MELEM),ULOCA(MNPEL), HEAT  30
     -          VLOCA(MNPEL),NDEST(MNPEL),LOCEL(MNPEL),LHEDV(MFRON), HEAT  31
     -          NADFM(MPOIN),NODFM(MPOIN),LNODS(MELEM,MNPEL),        HEAT  32
     -          PNORM(MFRON),GFLUM(MFRON,MFRON)                     HEAT  33
C                                                                 HEAT  34
C *** OPEN CHANNELS FOR READING AND WRITING                       HEAT  35
C                                                                 HEAT  36
      OPEN (UNIT=8,STATUS='OLD',FILE='INPUTS.DAT')               HEAT  37
      OPEN (UNIT=7,STATUS='NEW',FILE='OUTPUT.RES')              HEAT  38
C                                                                 HEAT  39
C *** READ IN DATA                                                HEAT  40
C                                                                 HEAT  41
      CALL READIN       (MELEM,MPOIN,MMATR,MNPEL,MDIME,MBOUN,MCDPT, HEAT  42
     -                   ITRAN,ILINR,IAXSY,ICONV,IPETR,IPHAS,NELEM, HEAT  43
     -                   NPOIN,NMATR,NNPEL,NFIXB,NEUMN,NTHER,NITER, HEAT  44
     -                   NTIML,TTIME,STIME,DTIME,DTMAX,NITUP,NITDN, HEAT  45
     -                   RELAX,TOLER,ALPHA,NTYPE,NCONC,NELMB,NFACB, HEAT  46
     -                   NFIXD,NLOAD,NCDPT,COORD,CONDY,CAPCY,TVALU, HEAT  47
     -                   CDVLU,CPVLU,TLIQS,TSOLS,QLATN,FIXED,FLUXE, HEAT  48
     -                   COEFF,RADIA,AMBIT,TLOAD,TEMPR,TILOD,VALOD, HEAT  49
     -                   UVELO,VVELO,FACTR)                        HEAT  50
C                                                                 HEAT  51
C *** CREATE REQUIRED INFORMATION                                 HEAT  52
C                                                                 HEAT  53
      CALL PREWRK       (MELEM,MPOIN,MNPEL,MBOUN,NELEM,NPOIN,NNPEL, HEAT  54
     -                   ICONV,ITRAN,NPROF,NCONC,NDIAG,NCOLM,NELMB, HEAT  55
     -                   NBELM,IFFIX,NFIXD,NEUMN,NNPFC,NFIXB,LNODS, HEAT  56
     -                   NADFM,NODFM,TFIXD,FIXED,ALPHA)            HEAT  57
C                                                                 HEAT  58
C *** IF SUPG METHOD IS TO BE USED FOR CONVECTION CALCULATE THE SCALAR HEAT  59
```

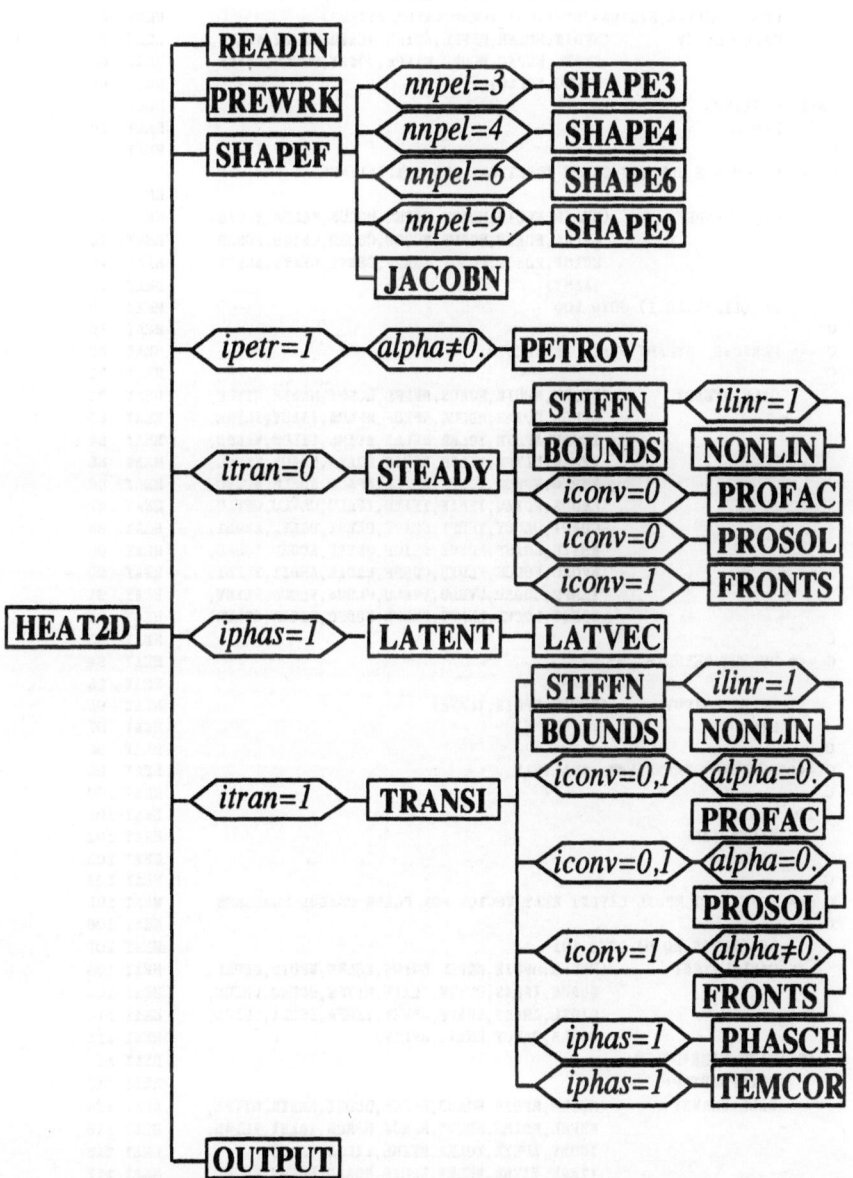

Figure A.1: HEAT2D Program structure

```
C *** COEFFICIENT FOR EACH ELEMENT                              HEAT  60
C                                                               HEAT  61
      IF (ICONV.EQ.1.AND.IPETR.EQ.1) THEN                       HEAT  62
      IF (ITRAN.EQ.1.AND.ALPHA.EQ.0.0) GOTO 311                 HEAT  63
      OPEN (UNIT=2,STATUS='SCRATCH',FORM='UNFORMATTED')         HEAT  64
      OPEN (UNIT=4,STATUS='SCRATCH',FORM='UNFORMATTED')         HEAT  65
      CALL PETROV        (NPOIN,NELEM,NNPEL,NDIME,NMATR,NPOIN,NELEM,  HEAT  66
     -                    NNPEL,NNPFC,NCONC,NTYPE,CONDY,COORD,PETRV,  HEAT  67
     -                    UVELO,VVELO)                          HEAT  68
  311 CONTINUE                                                  HEAT  69
      END IF                                                    HEAT  70
C                                                               HEAT  71
C *** CALCULATE ELEMENT SHAPE FUNCTIONS AND DERIVATIVES         HEAT  72
C                                                               HEAT  73
      CALL SHAPEF        (NELEM,NPOIN,NDIME,NNPEL,NGAUS,NELEM,NPOIN,   HEAT  74
     -                    NNPEL,NGASB,NGAUS,NCONC,COORD,WEIGB,POSGB,   HEAT  75
     -                    WEIGP,POSGX,POSGY,SHAPF,DERV1,DERV2,ARWET,   HEAT  76
     -                    IAXSY)                                HEAT  77
      IF (ITRAN.EQ.1) GOTO 100                                  HEAT  78
C                                                               HEAT  79
C *** PERFORM  STEADY STATE  ANALYSIS                           HEAT  80
C                                                               HEAT  81
      CALL STEADY        (NELEM,NPOIN,NGAUS,NNPEL,NCDPT,NMATR,NTYPE,   HEAT  82
     -                    NNPEL,NDIME,NBOUN,NPROF,NFRON,IAXSY,ILINR,   HEAT  83
     -                    ICONV,IPETR,TOLER,RELAX,NTIML,TILOD,VALOD,   HEAT  84
     -                    ITRAN,NITER,NELEM,NPOIN,NGASB,NGAUS,NNPEL,   HEAT  85
     -                    NNPFC,NPROF,NCDPT,NDFEL,NFACB,NBELM,NCONC,   HEAT  86
     -                    LNODS,NDIAG,IFFIX,TFIXD,TVALU,CDVLU,CPVLU,   HEAT  87
     -                    CONDY,CAPCY,TEMPT,SHAPF,DERV1,DERV2,ARWET,   HEAT  88
     -                    ASTIF,AMASS,POSGB,WEIGB,GSTIF,XCORD,YCORD,   HEAT  89
     -                    EFORC,FORCE,FLUXE,COEFF,RADIA,AMBIT,TLAST,   HEAT  90
     -                    TEMPR,COORD,UVELO,VVELO,ULOCA,VLOCA,PETRV,   HEAT  91
     -                    NDEST,LOCEL,LHEDV,NADFN,NODFN,PNORM,GFLUM)   HEAT  92
C                                                               HEAT  93
C *** OUTPUT RESULTS                                            HEAT  94
C                                                               HEAT  95
      CALL  OUTPUT       (NPOIN,NPOIN,TEMPR)                    HEAT  96
      STOP                                                      HEAT  97
C                                                               HEAT  98
C *** PERFORM  TRANSIENT  ANALYSIS                              HEAT  99
C                                                               HEAT 100
  100 CONTINUE                                                  HEAT 101
      TIME=STIME                                                HEAT 102
      NSTEP=0                                                   HEAT 103
C                                                               HEAT 104
C *** CONSTRUCT NODAL LATENT HEAT VECTOR FOR PHASE CHANGE PROBLEMS  HEAT 105
C                                                               HEAT 106
      IF (IPHAS.EQ.0) GOTO 101                                  HEAT 107
      CALL LATENT        (NELEM,NPOIN,NNPEL,NGAUS,NELEM,NPOIN,NNPEL,   HEAT 108
     -                    NGAUS,IPHAS,QLATN,ILATN,NTYPE,NCONC,VOLUN,   HEAT 109
     -                    QTOTL,AMASS,SHAPF,ARWET,TEMPR,ISOLI,TLIQS,   HEAT 110
     -                    TSOLS,CAPCY,NMATR,SPEAT)               HEAT 111
  101 TIME=TIME+DTIME                                           HEAT 112
      NSTEP=NSTEP+1                                             HEAT 113
      CALL TRANSI        (NELEM,NPOIN,NGAUS,NNPEL,NCDPT,NMATR,NTYPE,   HEAT 114
     -                    NNPEL,NDIME,NBOUN,NPROF,NFRON,IAXSY,ILINR,   HEAT 115
     -                    ICONV,IPETR,TOLER,NTIML,TILOD,VALOD,         HEAT 116
     -                    ITRAN,NITER,NELEM,NPOIN,NGASB,NGAUS,NNPEL,   HEAT 117
     -                    NNPFC,NPROF,NCDPT,NDFEL,NFACB,NBELM,NCONC,   HEAT 118
     -                    LNODS,NDIAG,IFFIX,TFIXD,TVALU,CDVLU,CPVLU,   HEAT 119
     -                    CONDY,CAPCY,TEMPT,SHAPF,DERV1,DERV2,ARWET,   HEAT 120
     -                    ASTIF,AMASS,POSGB,WEIGB,GSTIF,XCORD,YCORD,   HEAT 121
     -                    EFORC,FORCE,FLUXE,COEFF,RADIA,AMBIT,TLAST,   HEAT 122
     -                    TEMPR,COORD,UVELO,VVELO,ULOCA,VLOCA,PETRV,   HEAT 123
     -                    NDEST,LOCEL,LHEDV,NADFN,NODFN,PNORM,GFLUM,   HEAT 124
     -                    IPHAS,ILATN,NSTEP,ISOLI,LSOLI,             HEAT 125
     -                    NITUP,NITDN,TTIME,DTIME,DTMAX,ALPHA,FACTR,   HEAT 126
```

```
              -                TSOLS,TLIQS,SPEAT,TOLER,QLATH,TEMP1,VOLUM,       HEAT 127
              -                QTOTL,RVECT,QLATH,QCUMU,QIMCR,QRESI)             HEAT 128
C                                                                              HEAT 129
C *** OUTPUT RESULTS                                                           HEAT 130
C                                                                              HEAT 131
      WRITE (7,*) ' TIME = ',TIME, ' STEP = ',NSTEP                            HEAT 132
      CALL  OUTPUT    (NPOIN,NPOIN,TEMPR)                                      HEAT 133
      IF (TIME.LT.TTIME) GOTO 101                                             HEAT 134
      STOP                                                                     HEAT 135
      END                                                                      HEAT 136
```

A.4   Input Instructions

The input subroutine is listed here followed by complete instructions on preparing the input data file.

```
      SUBROUTINE READIN (NELEM,MPOIN,NMATR,NNPEL,NDIME,NBOUN,NCDPT,      READ  1
      -              ITRAN,ILINR,IAXSY,ICONV,IPETR,IPHAS,NELEM,          READ  2
      -              NPOIN,NMATR,NNPEL,NFIXB,NEUMN,NTHER,NITER,          READ  3
      -              NTIML,TTIME,STIME,DTIME,DTMAX,NITUP,NITDN,          READ  4
      -              RELAX,TOLER,ALPHA,NTYPE,NCONC,NELMB,NFACB,          READ  5
      -              NFIXD,NLOAD,NCDPT,COORD,CONDY,CAPCY,TVALU,          READ  6
      -              CDVLU,CPVLU,TLIQS,TSOLS,QLATH,FIXED,FLUXE,          READ  7
      -              COEFF,RADIA,AMBIT,TLOAD,TEMPR,TILOD,VALOD,          READ  8
      -              UVELO,VVELO,FACTR)                                  READ  9
C***************************************************************************   READ 10
C                                                                             READ 11
C**** THIS SUBROUTINE READS ALL PROBLEM DATA                                   READ 12
C                                                                             READ 13
C***************************************************************************   READ 14
C    INSERT DOUBLE                                                            READ 15
C                                                                             READ 16
      IMPLICIT REAL*8(A-H,O-Z)                                                READ 17
      DIMENSION NTYPE(NELEM),NCONC(NELEM,NNPEL),COORD(MPOIN,NDIME),           READ 18
      -         CONDY(NMATR),CAPCY(NMATR),FIXED(NBOUN),NFIXD(NBOUN),          READ 19
      -         NFACB(NBOUN),NELMB(NBOUN),NLOAD(NBOUN),FLUXE(NBOUN),          READ 20
      -         COEFF(NBOUN),RADIA(NBOUN),AMBIT(NBOUN),TLOAD(NBOUN),          READ 21
      -         TEMPR(MPOIN),CDVLU(NCDPT,NMATR),CPVLU(NCDPT,NMATR),           READ 22
      -         TVALU(NCDPT,NMATR),TILOD(NCDPT),VALOD(NCDPT),TITLE(12)        READ 23
      DIMENSION UVELO(MPOIN),VVELO(MPOIN)                                     READ 24
      READ(8,920)  TITLE                                                      READ 25
      WRITE(7,920) TITLE                                                      READ 26
  920 FORMAT(12A6)                                                            READ 27
      NITER=1                                                                 READ 28
C                                                                             READ 29
C *** 1. READ INFO FOR LINEAR/NONLINEAR/ STEADY STATE/TRANSIENT SOLUTION READ 30
C                                                                             READ 31
      READ (8,*) ITRAN,ILINR,IAXSY,ICONV,IPETR                               READ 32
      WRITE(7,901)ITRAN,ILINR,IAXSY,ICONV,IPETR                              READ 33
  901 FORMAT(//8H ITRAN =,I4,4X,8H ILINR =,I4,4X,8H IAXSY =,I4               READ 34
     .//8H ICONV =,I4,4X,8H IPETR =,I4)                                       READ 35
      IF (ITRAN.EQ.1) READ (8,*) TTIME,STIME,DTIME,ALPHA                      READ 36
      IF (ITRAN.EQ.1) WRITE(7,902)TTIME,STIME,DTIME,ALPHA                     READ 37
  902 FORMAT(//8H TTIME =,F5.3,3X,8H STIME =,F5.3,3X,                         READ 38
     .8H DTIME =,F5.3,3X,8H ALPHA =,F5.3)                                     READ 39
      IF (ILINR.EQ.1) READ (8,*) NITER,TOLER,RELAX                            READ 40
      IF (ILINR.EQ.1) WRITE(7,903) NITER,TOLER,RELAX                          READ 41
  903 FORMAT(//8H NITER =,I4,4X,8H TOLER =,F5.3,3X,                           READ 42
     .8H RELAX =,F5.3)                                                        READ 43
      IF (ILINR.EQ.1.AND.ITRAN.EQ.1) READ (8,*) NITUP,NITDN,DTMAX,FACTR READ 44
      IF (ILINR.EQ.1.AND.ITRAN.EQ.1) WRITE(7,907)NITUP,NITDN,DTMAX,FACTRREAD 45
  907 FORMAT(//8H NITUP =,I4,4X,8H NITDN =,I4,4X,8H DTMAX =,F5.3,3X,          READ 46
```

```
      .8H FACTR =,F5.3)                                             READ 47
C                                                                   READ 48
C *** 2. MESH DATA, CONNECTIVITY AND NODAL COORDINATES             READ 49
C                                                                   READ 50
      READ (8,*) NELEM,NPOIN,NMATR,NNPEL                            READ 51
      WRITE(7,904)NELEM,NPOIN,NMATR,NNPEL                           READ 52
  904 FORMAT(//8H NELEM =,I4,4X,8H NPOIN =,I4,4X,8H NMATR =,I4,4X,  READ 53
     .8H NNPEL =,I4//)                                              READ 54
      READ (8,*) ((NCONC(IEL,JPE),JPE=1,NNPEL),MTYPE(IEL),          READ 55
     -               IEL=1,NELEM)                                   READ 56
      WRITE(7,*)' ELEMENT, MATERIAL,      NODE NUMBERS'             READ 57
      DO 2 IELEM=1,NELEM                                            READ 58
    2 WRITE(7,905)IELEM,MTYPE(IELEM),(NCONC(IELEM,INPEL),INPEL=1,NNPEL) READ 59
  905 FORMAT(1X,I5,I9,6X,9I5)                                       READ 60
      READ (8,*) (COORD(INOD,1),COORD(INOD,2),INOD=1,NPOIN)         READ 61
      WRITE(7,*)                                                    READ 62
      WRITE(7,*)' NODE      X      Y'                               READ 63
      DO 10 IPOIN=1,NPOIN                                           READ 64
   10 WRITE(7,906) IPOIN,(COORD(IPOIN,IDIME),IDIME=1,2)             READ 65
  906 FORMAT(1X,I5,3F10.3)                                          READ 66
C                                                                   READ 67
C *** 3. MATERIAL PROPERTIES                                        READ 68
C                                                                   READ 69
      READ (8,*) (CONDY(IM),CAPCY(IM),IM=1,NMATR)                   READ 70
      WRITE(7,912)                                                  READ 71
  912 FORMAT(//'   MAT.',1X,'CONDUCTIVITY',1X,'CAPACITY')           READ 72
      DO 15 IM=1,NMATR                                              READ 73
   15 WRITE(7,906) IM,CONDY(IM),CAPCY(IM)                           READ 74
      IF (ILINR.EQ.1) THEN                                          READ 75
      READ (8,*) NCDPT                                              READ 76
      IF (NCDPT.GT.0) READ (8,*)                                    READ 77
     -  ((CDVLU(I,IM),CPVLU(I,IM),TVALU(I,IM),I=1,NCDPT),IM=1,NMATR) READ 78
      READ (8,*) IPHAS                                              READ 79
      IF (IPHAS.GT.0) READ (8,*) TLIQS,TSOLS,QLATN                  READ 80
      END IF                                                        READ 81
C                                                                   READ 82
C *** 4. DIRICHLET BOUNDARY CONDITIONS (FIXED TEMPERATURE)          READ 83
C                                                                   READ 84
      READ (8,*) NFIXB                                              READ 85
      IF (NFIXB.GT.0) READ (8,*) (NFIXD(IBC),FIXED(IBC),IBC=1,NFIXB) READ 86
      IF (NFIXB.GT.0) WRITE(7,*)                                    READ 87
      IF (NFIXB.GT.0) WRITE(7,*) ' NODE  FIXED TEMPR'               READ 88
      IF (NFIXB.GT.0) WRITE(7,916)(NFIXD(IBC),FIXED(IBC),IBC=1,NFIXB) READ 89
  916 FORMAT(I5,F13.4)                                              READ 90
C                                                                   READ 91
C *** 5. NEUMANN BOUNDARY CONDITIONS (FIXED FLUX, CONVECTION, RADIATIONREAD 92
C                                                                   READ 93
      READ (8,*) NEUMN                                              READ 94
      IF (NEUMN.GT.0) READ (8,*) (NELMB(IBC),NFACB(IBC),            READ 95
     -  FLUXE(IBC),COEFF(IBC),RADIA(IBC),AMBIT(IBC),IBC=1,NEUMN)    READ 96
      IF (NEUMN.GT.0) WRITE(7,*)                                    READ 97
      IF (NEUMN.GT.0)                                               READ 98
     -  WRITE(7,*) 'ELEMENT FACE  FLUX     COEFF     RADIA     AMBIT'READ 99
      IF (NEUMN.GT.0) WRITE(7,926)(NELMB(IBC),NFACB(IBC),           READ 100
     -  FLUXE(IBC),COEFF(IBC),RADIA(IBC),AMBIT(IBC),IBC=1,NEUMN)    READ 101
  926 FORMAT(2I5,4E11.3)                                            READ 102
C                                                                   READ 103
C *** 6. THERMAL LOADING (INTERNAL HEAT GENERATION)                 READ 104
C                                                                   READ 105
      READ (8,*) NTHER                                              READ 106
      IF (NTHER.GT.0) READ (8,*) (NLOAD(IBC),TLOAD(IBC),IBC=1,NTHER) READ 107
      IF (ITRAN.EQ.1.AND.NTHER.GT.0) THEN                           READ 108
      READ (8,*) NTIML                                              READ 109
      IF (NTIML.GT.0) READ (8,*) (TILOD(I),VALOD(I),I=1,NTIML)      READ 110
      END IF                                                        READ 111
C                                                                   READ 112
C *** 7. INITIAL TEMPERATURE FIELD FOR NONLINEAR OR TRANSIENT PROBLEMS READ 113
```

```
C                                                            READ 114
       IF (ILINR.EQ.1.OR.ITRAN.EQ.1) THEN                   READ 115
          READ (8,*) (TEMPR(INOD),INOD=1,NPOIN)             READ 116
C                                                            READ 117
C *** IMPOSE FIXED BOUNDARY CONDITIONS ON INITIAL TEMPERATURE FIELD  READ 118
C                                                            READ 119
          DO 11 I=1,NFIXB                                    READ 120
             TEMPR(NFIXD(I))=FIXED(I)                        READ 121
   11     CONTINUE                                           READ 122
          WRITE(7,*)                                         READ 123
          WRITE(7,*)'INITIAL TEMPERATURE'                    READ 124
          WRITE(7,910) (INOD,TEMPR(INOD),INOD=1,NPOIN)       READ 125
       END IF                                                READ 126
C                                                            READ 127
C *** 8.  VELOCITY FIELD FOR CONVECTION PROBLEMS             READ 128
C                                                            READ 129
       IF (ICONV.EQ.1) THEN                                  READ 130
          READ (8,*) (UVELO(INOD),VVELO(INOD),INOD=1,NPOIN)  READ 131
          WRITE(7,*)                                         READ 132
          WRITE(7,*)'SPECIFIED VELOCITY FIELD'               READ 133
          WRITE(7,*)' NODE      U-VEL      V-VEL'            READ 134
          DO 19 INOD=1,NPOIN                                 READ 135
             WRITE(7,911) INOD,UVELO(INOD),VVELO(INOD)       READ 136
   19     CONTINUE                                           READ 137
       END IF                                                READ 138
  910 FORMAT(4(I5,F10.4))                                    READ 139
  911 FORMAT(4(I5,F10.4,1X,F10.4))                           READ 140
       RETURN                                                READ 141
       END                                                   READ 142
```

Subroutine **READIN** reads in all the input data required by the program as listed above from the file **INPUTS.DAT**. The details of the input are described below. The variables to be input are written in bold face in bold brackets. All the variable names are as defined in the tables earlier.

1 Control data

    1.1 **(title)**

    1.2 **(itran,ilinr,iaxsy,iconv,ipetr)**

    1.3 **(ttime,stime,dtime,alpha)** (Only if *itran*=1)

    1.4 **(niter,toler,relax)** (Only if *ilinr*=1)

    1.5 **(nitup,nitdn,dtmax,factr)** (Only if *ilinr*=1 and *itran*=1)

2 Mesh data

    2.1 **(nelem,npoin,nmatr,nnpel)**

    2.2 **(nconc,mtype)** *nelem* times ('nnpel' nodes and material)

    2.3 **(coord)** *npoin* times (x-coordinate and y-coordinate)

3 Material properties

    3.1 **(condy,capcy)** *nmatr* times

    3.2 **(ncdpt)** (Only if *ilinr*=1)

3.3 **(cdvlu,cpvlu,tvalu)** *ncdpt* times *nmatr*times

The above set of data can be used for phase change problems using the enthalpy method. If *iphas*=0 (next line), and *itran*=1 and *ilinr*=1 then the program will expect an enthalpy curve to be input. An example of this appears in the documented examples section of Appendix B.

3.4 **(iphas)** (Only if *ilinr*=1)

If *iphas*=1 the source method will be used.

3.5 **(tliqs,tsols,qlatn)** (Only if *iphas*≠0)

4 Dirichlet type boundary conditions

4.1 **(nfixb)**

4.2 **(nfixd,fixed)** *nfixb* times

5 Neumann type boundary conditions

5.1 **(neumn)**

5.2 **(nelmb,nfacb,fluxe,coeff,radia,ambit)** *neumn* times

6 Thermal loading

6.1 **(nther)**

6.2 **(nload,tload)** *nther* times

6.3 **(ntiml)** (Only if *itran*=1)

6.4 **(tilod,valod)** *ntiml* times

7 Initial conditions

7.1 **(tempr)** *npoin* times (Only if *itran*=1 and *ilinr*=1)

8 Velocity field

8.1 **(uvelo,vvelo)** *npoin* times (Only if *iconv*=1)

A.5   Element Stiffness Calculations

Element stiffness and mass matrices are calculated in the following routine.

```
      SUBROUTINE STIFFN (NELEM,NGAUS,NNPEL,NCDPT,NMATR,IELEM,IMATR,      STIF  1
     -             NGAUS,NNPEL,ITRAN,ILINR,IPETR,ICONV,IAXSY,           STIF  2
     -             IPHAS,NCDPT,PETRO,TVALU,CDVLU,CPVLU,CONDT,           STIF  3
     -             CAPCT,TEMPT,SHAPF,DERV1,DERV2,ARWET,ASTIF,           STIF  4
     -             AMASS,ULOCA,VLOCA,XCORD,YCORD)                       STIF  5
C***************************************************************************  STIF  6
C                                                                        STIF  7
C**** THIS SUBROUTINE CONSTRUCTS ELEMENT STIFFNESS MATRIX                STIF  8
```

```
C                                                                        STIF   9
C************************************************************************ STIF  10
C     INSERT DOUBLE                                                       STIF  11
C                                                                        STIF  12
      IMPLICIT REAL*8(A-H,O-Z)                                           STIF  13
      DIMENSION SHAPF(MNPEL,MGAUS,MELEM),DERV1(MNPEL,MGAUS,MELEM),       STIF  14
     -          DERV2(MNPEL,MGAUS,MELEM),ARWET(MGAUS,MELEM),             STIF  15
     -          CDVLU(MCDPT,MMATR),CPVLU(MCDPT,MMATR),                   STIF  16
     -          TVALU(MCDPT,MMATR),TEMPT(MNPEL),ASTIF(MNPEL,MNPEL),      STIF  17
     -          AMASS(MNPEL,MNPEL),ULOCA(MNPEL),VLOCA(MNPEL),            STIF  18
     -          XCORD(MNPEL),YCORD(MNPEL),SHAPP(9),CARTP(2,9)            STIF  19
C                                                                        STIF  20
C *** EXTRACT ELEMENT SHAPE FUNCTIONS                                     STIF  21
C                                                                        STIF  22
      DO 100 IGAUS=1,NGAUS                                               STIF  23
         DAREA=ARWET(IGAUS,IELEM)                                        STIF  24
         DO 30 INODP=1,NNPEL                                             STIF  25
            SHAPP(INODP)=SHAPF(INODP,IGAUS,IELEM)                        STIF  26
            CARTP(1,INODP)=DERV1(INODP,IGAUS,IELEM)                      STIF  27
            CARTP(2,INODP)=DERV2(INODP,IGAUS,IELEM)                      STIF  28
   30    CONTINUE                                                        STIF  29
C                                                                        STIF  30
C *** EVALUATE PROPERTIES AT INTEGRATION POINTS                           STIF  31
C                                                                        STIF  32
         TEMP1=0.0                                                       STIF  33
         TEMP2=0.0                                                       STIF  34
         TEMP3=0.0                                                       STIF  35
         COND=CONDT                                                      STIF  36
         CAPY=CAPCT                                                      STIF  37
         RADI=1.0D0                                                      STIF  38
         IF (IAXSY.EQ.1) THEN                                            STIF  39
            RADI=0.0D0                                                   STIF  40
            DO 49 INODP=1,NNPEL                                          STIF  41
               RADI=RADI+XCORD(INODP)*SHAPP(INODP)                       STIF  42
   49       CONTINUE                                                     STIF  43
         END IF                                                          STIF  44
         IF (ILINR.EQ.1) THEN                                            STIF  45
            TEMP =0.0                                                    STIF  46
            DTX=0.0                                                      STIF  47
            DTY=0.0                                                      STIF  48
            DHX=0.0                                                      STIF  49
            DHY=0.0                                                      STIF  50
            DO 50 INODP=1,NNPEL                                          STIF  51
               TEMT=TEMPT(INODP)                                        STIF  52
               TEMP=TEMP+TEMT*SHAPP(INODP)                              STIF  53
               IF (ITRAN.EQ.1.AND.IPHAS.EQ.0) THEN                      STIF  54
               CALL NONLIN (MCDPT,MMATR,IMATR,NCDPT,TEMT,TVALU,CPVLU,VALUE)STIF  55
                  ENTH=VALUE                                            STIF  56
                  DTX=DTX+TEMT*CARTP(1,INODP)                           STIF  57
                  DTY=DTY+TEMT*CARTP(2,INODP)                           STIF  58
                  DHX=DHX+ENTH*CARTP(1,INODP)                           STIF  59
                  DHY=DHY+ENTH*CARTP(2,INODP)                           STIF  60
               END IF                                                   STIF  61
   50       CONTINUE                                                     STIF  62
            CALL NONLIN (MCDPT,MMATR,IMATR,NCDPT,TEMP,TVALU,CDVLU,VALUE)STIF  63
            COND=COND*VALUE                                              STIF  64
            IF (ITRAN.EQ.1.AND.IPHAS.EQ.0) THEN                         STIF  65
               IF (DTX.NE.0.0.AND.DTY.NE.0.0) THEN                      STIF  66
C *** --------------- MORGAN ET. AL. AND LEMMON'S APPROX. -------------STIF  67
               CAPY=CAPY*DSQRT((DHX*DHX+DHY*DHY)/(DTX*DTX+DTY*DTY))      STIF  68
C              IF (CAPY.LE.0.0) CAPY=CAPCT                              STIF  69
C *** ----------------- DELUIDICE ET. AL.'S  APPROX. --------------------STIF  70
C              CAPY=CAPY*(DHX*DTX+DHY*DTY)/(DTX*DTX+DTY*DTY)            STIF  71
               IF (CAPY.LE.0.0) CAPY=CAPCT                              STIF  72
               END IF                                                   STIF  73
            ELSE IF (ITRAN.NE.1) THEN                                   STIF  74
               CALL NONLIN (MCDPT,MMATR,IMATR,NCDPT,TEMP,TVALU,CPVLU,   STIF  75
```

```
         -                   VALUE)                          STIF  76
                    CAPY=CAPY*VALUE                          STIF  77
               END IF                                       STIF  78
            END IF                                          STIF  79
            IF (ICONV.EQ.1) THEN                            STIF  80
               UVEL =0.0                                    STIF  81
               VVEL =0.0                                    STIF  82
               DO 51 INODP=1,NNPEL                          STIF  83
                  UVEL=UVEL+ULOCA(INODP)*SHAPP(INODP)       STIF  84
                  VVEL=VVEL+VLOCA(INODP)*SHAPP(INODP)       STIF  85
     51        CONTINUE                                     STIF  86
            END IF                                          STIF  87
C                                                           STIF  88
C *** CONSTRUCT STIFFNESS AND MASS MATRICES                 STIF  89
C                                                           STIF  90
            DO 90 INODP=1,NNPEL                             STIF  91
               SHAPI=SHAPP(INODP)                           STIF  92
               CARXI=CARTP(1,INODP)                         STIF  93
               CARYI=CARTP(2,INODP)                         STIF  94
               DO 80 JNODP=1,NNPEL                          STIF  95
                  SHAPJ=SHAPP(JNODP)                        STIF  96
                  CARXJ=CARTP(1,JNODP)                      STIF  97
                  CARYJ=CARTP(2,JNODP)                      STIF  98
C                                                           STIF  99
C *** DIFFUSION PART OF STIFFNESS MATRIX.                   STIF 100
C                                                           STIF 101
                  TEMP1=(CARXI*CARXJ+CARYI*CARYJ)*COND*DAREA*RADI  STIF 102
C                                                           STIF 103
C *** CONVECTION PART OF STIFFNESS MATRIX.                  STIF 104
C                                                           STIF 105
                  IF (ICONV.EQ.1)                           STIF 106
         -        TEMP2=(UVEL*CARXJ+VVEL*CARYJ)*SHAPI*DAREA*CAPY*RADI  STIF 107
C                                                           STIF 108
C *** PETROV-GALERKIN PART OF STIFFNESS MATRIX.             STIF 109
C                                                           STIF 110
                  IF (IPETR.EQ.1)                           STIF 111
         -        TEMP3=(CARXI*CARXJ*UVEL*UVEL+CARYI*CARYJ*VVEL*VVEL+  STIF 112
         -               CARXI*CARYJ*UVEL*VVEL+CARYI*CARXJ*VVEL*UVEL)  STIF 113
         -              *PETRO*DAREA*CAPY*RADI              STIF 114
                  ASTIF(INODP,JNODP)=ASTIF(INODP,JNODP)+TEMP1+TEMP2+TEMP3  STIF 115
C                                                           STIF 116
C *** CAPACITY MATRIX FOR TRANSIENT PROBLEMS                STIF 117
C                                                           STIF 118
                  IF (ITRAN.EQ.1)                           STIF 119
         -        AMASS(INODP,JNODP)=AMASS(INODP,JNODP)+     STIF 120
         -                     SHAPI*SHAPJ*CAPY*DAREA       STIF 121
     80        CONTINUE                                     STIF 122
     90     CONTINUE                                        STIF 123
    100 CONTINUE                                            STIF 124
        RETURN                                              STIF 125
        END                                                 STIF 126
```

A.6   Phase Change Calculations

Phase change calculations are performed in the following routines.

A.6.1   Nodal Latent Heat Calculations

```
        SUBROUTINE LATENT (NELEM,MPOIN,MNPEL,MGAUS,NELEM,NPOIN,NNPEL,   LATE  1
         -                 NGAUS,IPHAS,QLATN,ILATN,MTYPE,NCONC,VOLUM,   LATE  2
         -                 QTOTL,AMASS,SHAPF,ARWET,TEMPR,ISOLI,TLIQS,   LATE  3
         -                 TSOLS,CAPCY,MMATR,SPEAT)                     LATE  4
```

```
C**********************************************************************   LATE   5
C                                                                        LATE   6
C**** THIS SUBROUTINE DEALS WITH LATENT HEAT                             LATE   7
C                                                                        LATE   8
C**********************************************************************   LATE   9
C     INSERT DOUBLE                                                      LATE  10
C                                                                        LATE  11
      IMPLICIT REAL*8(A-H,O-Z)                                           LATE  12
      DIMENSION NCONC(MELEM,MNPEL),ILATN(MPOIN),MTYPE(MELEM),            LATE  13
     -          AMASS(MNPEL,MNPEL),VOLUM(MPOIN),QTOTL(MPOIN),            LATE  14
     -          ARWET(MGAUS,MELEM),SHAPF(MNPEL,MGAUS,MELEM),             LATE  15
     -          TEMPR(MPOIN),ISOLI(MPOIN),CAPCY(MMATR)                   LATE  16
C                                                                        LATE  17
C *** CONSTRUCT NODAL LATENT HEAT VECTOR FOR PHASE CHANGE PROBLEMS       LATE  18
C                                                                        LATE  19
      DO 134 I=1,NPOIN                                                   LATE  20
      ILATN(I)=0                                                         LATE  21
      IF (TEMPR(I).GT.TLIQS) THEN                                        LATE  22
         ISOLI(I)=-1                                                     LATE  23
        ELSE IF (TEMPR(I).LT.TSOLS) THEN                                 LATE  24
         ISOLI(I)=1                                                      LATE  25
        ELSE                                                             LATE  26
         ISOLI(I)=0                                                      LATE  27
      END IF                                                             LATE  28
 134  CONTINUE                                                           LATE  29
      CALL LATVEC     (MELEM,MPOIN,MNPEL,MGAUS,NELEM,NPOIN,NNPEL,        LATE  30
     -                 NGAUS,IPHAS,QLATN,ILATN,MTYPE,NCONC,VOLUM,        LATE  31
     -                 QTOTL,AMASS,SHAPF,ARWET)                          LATE  32
C                                                                        LATE  33
      SPEAT=CAPCY(IPHAS)                                                 LATE  34
      RETURN                                                             LATE  35
      END                                                                LATE  36

      SUBROUTINE LATVEC (MELEM,MPOIN,MNPEL,MGAUS,NELEM,NPOIN,NNPEL,      LATV   1
     -                   NGAUS,IPHAS,QLATN,ILATN,MTYPE,NCONC,VOLUM,      LATV   2
     -                   QTOTL,AMASS,SHAPF,ARWET)                        LATV   3
C**********************************************************************   LATV   4
C                                                                        LATV   5
C**** THIS SUBROUTINE CONSTRUCTS THE GLOBAL NODAL LATENT HEAT VECTOR     LATV   6
C                                                                        LATV   7
C**********************************************************************   LATV   8
C     INSERT DOUBLE                                                      LATV   9
C                                                                        LATV  10
      IMPLICIT REAL*8(A-H,O-Z)                                           LATV  11
      DIMENSION NCONC(MELEM,MNPEL),ILATN(MPOIN),MTYPE(MELEM),            LATV  12
     -          AMASS(MNPEL,MNPEL),VOLUM(MPOIN),QTOTL(MPOIN),            LATV  13
     -          ARWET(MGAUS,MELEM),SHAPF(MNPEL,MGAUS,MELEM),SHAPE(9)     LATV  14
      DO 1000 IEL=1,NELEM                                                LATV  15
      IMAT=MTYPE(IEL)                                                    LATV  16
      IF (IMAT.NE.IPHAS) GOTO 1000                                       LATV  17
      DO 1008 I=1,NNPEL                                                  LATV  18
         DO 1009 J=1,NNPEL                                               LATV  19
         AMASS(I,J)=0.0                                                  LATV  20
 1009    CONTINUE                                                        LATV  21
 1008  CONTINUE                                                          LATV  22
      DO 1001 IGAUS=1,NGAUS                                              LATV  23
         DO 1002 I=1,NNPEL                                               LATV  24
         SHAPE(I)=SHAPF(I,IGAUS,IEL)                                     LATV  25
 1002    CONTINUE                                                        LATV  26
         GWIGJ=ARWET(IGAUS,IEL)                                          LATV  27
         DO 1003 I=1,NNPEL                                               LATV  28
            DO 1004 J=1,NNPEL                                            LATV  29
            AMASS(I,J)=AMASS(I,J)+SHAPE(I)*SHAPE(J)*GWIGJ                LATV  30
 1004       CONTINUE                                                     LATV  31
 1003    CONTINUE                                                        LATV  32
 1001  CONTINUE                                                          LATV  33
      DO 1005 I=1,NNPEL                                                  LATV  34
```

```
                  NOD=NCONC(IEL,I)                                    LATV  35
                  DO 1006 J=1,NNPEL                                   LATV  36
                     ILATN(NOD)=1                                     LATV  37
                     VOLUM(NOD)=VOLUM(NOD)+AMASS(I,J)                 LATV  38
1006           CONTINUE                                              LATV  39
1005      CONTINUE                                                   LATV  40
1000 CONTINUE                                                        LATV  41
        DO 1010 I=1,NPOIN                                            LATV  42
           QTOTL(I)=VOLUM(I)*QLATN                                   LATV  43
1010 CONTINUE                                                        LATV  44
     RETURN                                                          LATV  45
     END                                                            LATV  46
```

A.6.2   Latent Heat Release for Each Iteration

```
      SUBROUTINE PHASCH (MPOIN,MMATR,MCDPT,NPOIN,NCDPT,NSTEP,IPHAS,   PHAS   1
     -                   TLIQS,TSOLS,SPEAT,QLATN,ISOLI,LSOLI,ILATN,  PHAS   2
     -                   TEMPR,TLAST,VOLUM,QTOTL,QCUMU,QLATH,QINCR,   PHAS   3
     -                   QRESI,CPVLU,TVALU)                          PHAS   4
C*******************************************************************  PHAS   5
C                                                                   PHAS   6
C**** THIS SUBROUTINE CALCULATES THE LATENT HEAT SOURCE TERM FOR EACH PHAS   7
C**** ITERATION.                                                    PHAS   8
C                                                                   PHAS   9
C*******************************************************************  PHAS  10
C     INSERT DOUBLE                                                 PHAS  11
C                                                                   PHAS  12
      IMPLICIT REAL*8(A-H,O-Z)                                      PHAS  13
      DIMENSION ISOLI(MPOIN),LSOLI(MPOIN),ILATN(MPOIN),TEMPR(MPOIN), PHAS  14
     -          TLAST(MPOIN),VOLUM(MPOIN),QTOTL(MPOIN),QCUMU(MPOIN), PHAS  15
     -          QLATH(MPOIN),QINCR(MPOIN),CPVLU(MCDPT,MMATR),        PHAS  16
     -          QRESI(MPOIN),TVALU(MCDPT,MMATR)                     PHAS  17
C                                                                   PHAS  18
      DO 1000 I=1,NPOIN                                             PHAS  19
         IF (ILATN(I).EQ.0) GOTO 1000                               PHAS  20
         TEMP=TEMPR(I)                                              PHAS  21
         CALL NONLIN (MCDPT,MMATR,IPHAS,NCDPT,TEMP,TVALU,CPVLU,VALUE) PHAS  22
         STEMP=SPEAT*VALUE                                          PHAS  23
         SHEAT=1.0/((TLIQS-TSOLS)/QLATN+1.0/STEMP)                  PHAS  24
         QINCR(I)=0.0                                               PHAS  25
         IF (ISOLI(I).EQ.-1) THEN                                   PHAS  26
            LSOLI(I)=ISOLI(I)*NSTEP                                 PHAS  27
            IF (TEMPR(I).LT.TLIQS.AND.TLAST(I).GE.TLIQS) ISOLI(I)=0 PHAS  28
            IF (ISOLI(I).NE.0) GOTO 1000                            PHAS  29
            DTEMP=TLIQS-TEMPR(I)                                    PHAS  30
            QINCR(I)=VOLUM(I)*SHEAT*DTEMP                           PHAS  31
            IF (QINCR(I).GT.QTOTL(I)) QINCR(I)=QTOTL(I)             PHAS  32
            QCUMU(I)=QCUMU(I)+QINCR(I)                              PHAS  33
            QLATH(I)=QLATH(I)+QINCR(I)                              PHAS  34
            GOTO 1000                                               PHAS  35
         END IF                                                     PHAS  36
         IF (ISOLI(I).EQ.1) THEN                                    PHAS  37
            LSOLI(I)=ISOLI(I)*NSTEP                                 PHAS  38
            IF (TEMPR(I).GT.TSOLS.AND.TLAST(I).LE.TSOLS) ISOLI(I)=0 PHAS  39
            IF (ISOLI(I).NE.0) GOTO 1000                            PHAS  40
            DTEMP=TSOLS-TEMPR(I)                                    PHAS  41
            QINCR(I)=VOLUM(I)*SHEAT*DTEMP                           PHAS  42
            IF (DABS(QINCR(I)).GT.QTOTL(I)) QINCR(I)=-QTOTL(I)      PHAS  43
            QCUMU(I)=QCUMU(I)+QINCR(I)                              PHAS  44
            QLATH(I)=QLATH(I)+QINCR(I)                              PHAS  45
            GOTO 1000                                               PHAS  46
         END IF                                                     PHAS  47
         IF (ISOLI(I).EQ.0) THEN                                    PHAS  48
            IF (NSTEP.EQ.IABS(LSOLI(I))) THEN                       PHAS  49
```

```
      IF (LSOLI(I).LT.0) DTEMP=TLIQS-TEMPR(I)            PHAS  50
      IF (LSOLI(I).GT.0) DTEMP=TSOLS-TEMPR(I)            PHAS  51
    ELSE                                                 PHAS  52
      DTEMP=TLAST(I)-TEMPR(I)                            PHAS  53
    END IF                                               PHAS  54
    QINCR(I)=VOLUM(I)*SHEAT*DTEMP                        PHAS  55
    QTEMP=QCUMU(I)                                       PHAS  56
    QCUMU(I)=QCUMU(I)+QINCR(I)                           PHAS  57
    QLATH(I)=QLATH(I)+QINCR(I)                           PHAS  58
    IF (DTEMP.LT.0) THEN                                 PHAS  59
      IF (DABS(QCUMU(I)).GT.QTOTL(I)) THEN               PHAS  60
        QRESI(I)=-QTOTL(I)-QTEMP                         PHAS  61
        QLATH(I)=QLATH(I)-QCUMU(I)-QTOTL(I)              PHAS  62
        QCUMU(I)=-QTOTL(I)                               PHAS  63
      END IF                                             PHAS  64
    ELSE                                                 PHAS  65
      IF (DABS(QCUMU(I)).GT.QTOTL(I)) THEN               PHAS  66
        QRESI(I)=QTOTL(I)-QTEMP                          PHAS  67
        QLATH(I)=QLATH(I)-QCUMU(I)+QTOTL(I)              PHAS  68
        QCUMU(I)=QTOTL(I)                                PHAS  69
      END IF                                             PHAS  70
    END IF                                               PHAS  71
    END IF                                               PHAS  72
1000 CONTINUE                                            PHAS  73
    RETURN                                               PHAS  74
    END                                                  PHAS  75
```

A.6.3 Correction of Temperatures after Each Iteration

```
      SUBROUTINE TEMCOR (NPOIN,MPOIN,TLIQS,TSOLS,ISOLI,LSOLI,ILATN,    TEMC   1
    -                   TEMPR,TLAST,VOLUM,QTOTL,QCUMU)                 TEMC   2
C**************************************************************************  TEMC   3
C                                                                     TEMC   4
C**** SUBROUTINE TO CORRECT NODAL TEMPERATURES FOR CONSISTENCY WITH   TEMC   5
C**** HEAT EVOLVED.                                                   TEMC   6
C                                                                     TEMC   7
C**************************************************************************  TEMC   8
C   INSERT DOUBLE                                                     TEMC   9
C                                                                     TEMC  10
      IMPLICIT REAL*8(A-H,O-Z)                                        TEMC  11
      DIMENSION ISOLI(MPOIN),LSOLI(MPOIN),ILATN(MPOIN),TEMPR(MPOIN),  TEMC  12
    -           TLAST(MPOIN),VOLUM(MPOIN),QTOTL(MPOIN),QCUMU(MPOIN)   TEMC  13
      TRANG=TLIQS-TSOLS                                               TEMC  14
      DO 1000 I=1,NPOIN                                               TEMC  15
        IF (ILATN(I).EQ.0) GOTO 1000                                 TEMC  16
        IF (ISOLI(I).NE.0) GOTO 1000                                 TEMC  17
        IF (TRANG.EQ.0) THEN                                         TEMC  18
          TEMPR(I)=TLIQS                                             TEMC  19
          GOTO 1001                                                 TEMC  20
        END IF                                                       TEMC  21
        IF (LSOLI(I).LT.0) THEN                                      TEMC  22
          TEMP=TLIQS-(DABS(QCUMU(I))/QTOTL(I))*TRANG                 TEMC  23
          TEMPR(I)=TEMP                                              TEMC  24
        ELSE IF (LSOLI(I).GT.0) THEN                                 TEMC  25
          TEMP=TSOLS+(DABS(QCUMU(I))/QTOTL(I))*TRANG                 TEMC  26
          TEMPR(I)=TEMP                                              TEMC  27
        END IF                                                       TEMC  28
1001    IF (QCUMU(I).LT.0..AND.DABS(QCUMU(I)).GE.QTOTL(I)) ISOLI(I)=-1 TEMC  29
        IF (QCUMU(I).GT.0.0.AND.QCUMU(I).GE.QTOTL(I)) ISOLI(I)=1     TEMC  30
1000 CONTINUE                                                        TEMC  31
    RETURN                                                           TEMC  32
    END                                                              TEMC  33
```

A.7   Documented Examples

Details of two examples are listed in the following sections. The first one is a transient heat conduction problem, the second is a steady state forced convection problem.

A.7.1   1-D Solidification Example

The example chosen is that of a 1-D rod of material subjected to a fixed temperature at one end, insulated at the other, undergoing solidification as it cools beginning at the fixed end. This example requires a transient and non-linear analysis. Figure A.2 shows the geometry, boundary conditions, initial conditions and the material properties. The analysis is carried on to a time of 4.0 seconds. The analytical solution to this problem is given in [3], according to which the temperature at $x = 1.0cm$ is $-22.63°C$. A mesh of 20 9-noded elements is used to solve this example. Graphical results for a single node point (node 21) for this problem appeared earlier in Chapter 5.

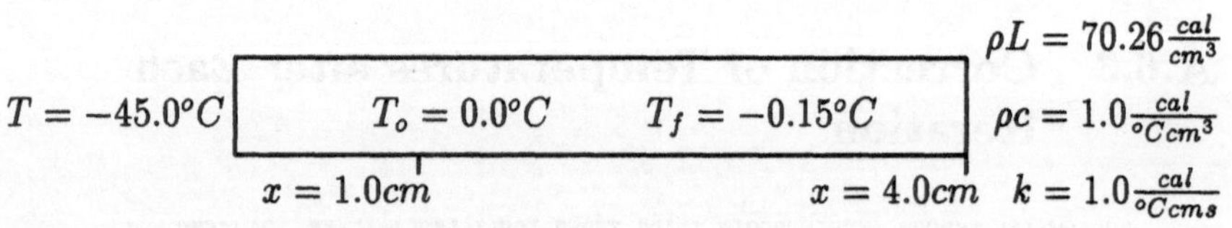

$$T = -45.0°C \quad \boxed{\qquad T_o = 0.0°C \qquad T_f = -0.15°C \qquad} \quad \begin{array}{l} \rho L = 70.26 \frac{cal}{cm^3} \\ \rho c = 1.0 \frac{cal}{°C\,cm^3} \\ k = 1.0 \frac{cal}{°C\,cm\,s} \end{array}$$

$$x = 1.0cm \qquad\qquad\qquad\qquad x = 4.0cm$$

Figure A.2: Solidification example.

Input Data File

Contents of the **INPUTS.DAT** for this example are as follows:

```
1-D Solidification
   1  1  0  0  0
   4.0  0.0  0.1  1.0
   9   0.01  1.0
   2   4   0.1  2.0
  20  123  1  9
  39  40  41  82  123  122  121  80  81   1
  37  38  39  80  121  120  119  78  79   1
  35  36  37  78  119  118  117  76  77   1
  33  34  35  76  117  116  115  74  75   1
  31  32  33  74  115  114  113  72  73   1
  29  30  31  72  113  112  111  70  71   1
  27  28  29  70  111  110  109  68  69   1
  25  26  27  68  109  108  107  66  67   1
  23  24  25  66  107  106  105  64  65   1
  21  22  23  64  105  104  103  62  63   1
  19  20  21  62  103  102  101  60  61   1
  17  18  19  60  101  100   99  58  59   1
  15  16  17  58   99   98   97  56  57   1
```

```
13  14  15  56  97  96  95  54  55   1
11  12  13  54  95  94  93  52  53   1
 9  10  11  52  93  92  91  50  51   1
 7   8   9  50  91  90  89  48  49   1
 5   6   7  48  89  88  87  46  47   1
 3   4   5  46  87  86  85  44  45   1
 1   2   3  44  85  84  83  42  43   1
0.  0.
    4.87805E-03  0.
    9.75610E-03  0.
    2.46470E-02  0.
    3.95379E-02  0.
    6.44416E-02  0.
    8.93453E-02  0.
   0.124262  0.
   0.159178  0.
   0.204108  0.
   0.249037  0.
   0.303979  0.
   0.358922  0.
   0.423877  0.
   0.488832  0.
   0.563800  0.
   0.638768  0.
   0.723748  0.
   0.808729  0.
   0.904365  0.
    1.00000  0.
    1.10436  0.
    1.20873  0.
    1.32375  0.
    1.43877  0.
    1.56380  0.
    1.68883  0.
    1.82388  0.
    1.95892  0.
    2.10398  0.
    2.24904  0.
    2.40411  0.
    2.55918  0.
    2.72426  0.
    2.88935  0.
    3.06444  0.
    3.23954  0.
    3.42465  0.
    3.60976  0.
    3.80488  0.
    4.00000  0.
0.      0.250000
    4.87805E-03  0.250000
    9.75610E-03  0.250000
    2.46470E-02  0.250000
    3.95379E-02  0.250000
    6.44416E-02  0.250000
    8.93453E-02  0.250000
   0.124262  0.250000
   0.159178  0.250000
   0.204108  0.250000
   0.249037  0.250000
   0.303979  0.250000
   0.358922  0.250000
   0.423877  0.250000
   0.488832  0.250000
   0.563800  0.250000
   0.638768  0.250000
   0.723748  0.250000
   0.808729  0.250000
```

```
     0.904365   0.250000
     1.00000    0.250000
     1.10436    0.250000
     1.20873    0.250000
     1.32375    0.250000
     1.43877    0.250000
     1.56380    0.250000
     1.68883    0.250000
     1.82388    0.250000
     1.95892    0.250000
     2.10398    0.250000
     2.24904    0.250000
     2.40411    0.250000
     2.55918    0.250000
     2.72426    0.250000
     2.88935    0.250000
     3.06444    0.250000
     3.23954    0.250000
     3.42465    0.250000
     3.60976    0.250000
     3.80488    0.250000
     4.00000    0.250000
0.   0.500000
     4.87805E-03   0.500000
     9.75610E-03   0.500000
     2.46470E-02   0.500000
     3.95379E-02   0.500000
     6.44416E-02   0.500000
     8.93453E-02   0.500000
     0.124262   0.500000
     0.159178   0.500000
     0.204108   0.500000
     0.249037   0.500000
     0.303979   0.500000
     0.358922   0.500000
     0.423877   0.500000
     0.488832   0.500000
     0.563800   0.500000
     0.638768   0.500000
     0.723748   0.500000
     0.808729   0.500000
     0.904365   0.500000
     1.00000    0.500000
     1.10436    0.500000
     1.20873    0.500000
     1.32375    0.500000
     1.43877    0.500000
     1.56380    0.500000
     1.68883    0.500000
     1.82388    0.500000
     1.95892    0.500000
     2.10398    0.500000
     2.24904    0.500000
     2.40411    0.500000
     2.55918    0.500000
     2.72426    0.500000
     2.88935    0.500000
     3.06444    0.500000
     3.23954    0.500000
     3.42465    0.500000
     3.60976    0.500000
     3.80488    0.500000
     4.00000    0.500000
     1.08  1.0
     0
     1
     -0.15  -0.15  70.26
```

```
3
1   -45.0
83  -45.0
42  -45.0
0
0
0.0
 .
 .
 .
0.0
```

The last set of data consists of *npoin* lines of initial temperatures which are all the same therefore only two lines are shown.

Output Data File

The output from **HEAT2D** is written into the file **OUTPUT.RES**. This file includes some information from the input data file. Temperature results for only the last timestep are listed.

```
1-D Solidification


ITRAN =   1     ILINR =   1     IAXSY =   0

ICONV =   0     IPETR =   0


TTIME =4.000   STIME =0.000   DTIME =0.100   ALPHA =1.000


NITER =   9     TOLER =0.01    RELAX =1.000


NITUP =   2     NITDN =   4     DTMAX =0.100   FACTR =2.000


NELEM =  20     NPOIN = 123    NMATR =   1     NNPEL =   9


ELEMENT, MATERIAL,    NODE NUMBERS
  1       1         39   40   41   82  123  122  121   80   81
```

element information as in input data

```
 20       1          1    2    3   44   85   84   83   42   43

NODE       X        Y
  1      0.000    0.000
```

nodal coordinates as in input data

```
123      4.000    0.500


    MAT. CONDUCTIVITY CAPACITY
    1     1.080      1.000

NODE  FIXED TEMPR
  1     -45.0000
 83     -45.0000
 42     -45.0000
```

INITIAL TEMPERATURE

1	-45.0000	2	0.0000	3	0.0000	4	0.0000
5	0.0000	6	0.0000	7	0.0000	8	0.0000
9	0.0000	10	0.0000	11	0.0000	12	0.0000
13	0.0000	14	0.0000	15	0.0000	16	0.0000
17	0.0000	18	0.0000	19	0.0000	20	0.0000
21	0.0000	22	0.0000	23	0.0000	24	0.0000
25	0.0000	26	0.0000	27	0.0000	28	0.0000
29	0.0000	30	0.0000	31	0.0000	32	0.0000
33	0.0000	34	0.0000	35	0.0000	36	0.0000
37	0.0000	38	0.0000	39	0.0000	40	0.0000
41	0.0000	42	-45.0000	43	0.0000	44	0.0000
45	0.0000	46	0.0000	47	0.0000	48	0.0000
49	0.0000	50	0.0000	51	0.0000	52	0.0000
53	0.0000	54	0.0000	55	0.0000	56	0.0000
57	0.0000	58	0.0000	59	0.0000	60	0.0000
61	0.0000	62	0.0000	63	0.0000	64	0.0000
65	0.0000	66	0.0000	67	0.0000	68	0.0000
69	0.0000	70	0.0000	71	0.0000	72	0.0000
73	0.0000	74	0.0000	75	0.0000	76	0.0000
77	0.0000	78	0.0000	79	0.0000	80	0.0000
81	0.0000	82	0.0000	83	-45.0000	84	0.0000
85	0.0000	86	0.0000	87	0.0000	88	0.0000
89	0.0000	90	0.0000	91	0.0000	92	0.0000
93	0.0000	94	0.0000	95	0.0000	96	0.0000
97	0.0000	98	0.0000	99	0.0000	100	0.0000
101	0.0000	102	0.0000	103	0.0000	104	0.0000
105	0.0000	106	0.0000	107	0.0000	108	0.0000
109	0.0000	110	0.0000	111	0.0000	112	0.0000
113	0.0000	114	0.0000	115	0.0000	116	0.0000
117	0.0000	118	0.0000	119	0.0000	120	0.0000
121	0.0000	122	0.0000	123	0.0000		

SOLUTION CONVERGED AT ITERATION NO.   2
TIME =   4.000000000000003   STEP =   47

*** OUTPUT OF TEMPERATURE ***

1	-45.0000	2	-44.8883	3	-44.7766	4	-44.4357
5	-44.0947	6	-43.5246	7	-42.9547	8	-42.1558
9	-41.3575	10	-40.3311	11	-39.3060	12	-38.0547
13	-36.8062	14	-35.3346	15	-33.8686	16	-32.1845
17	-30.5100	18	-28.6251	19	-26.7554	20	-24.6715
21	-22.6106	22	-20.3893	23	-18.1977	24	-15.8174
25	-13.4729	26	-10.9621	27	-8.4859	28	-5.8172
29	-0.1500	30	-0.1500	31	-0.1500	32	-0.1500
33	-0.1500	34	-0.1500	35	-0.1500	36	-0.1500
37	0.0166	38	0.0171	39	0.0177	40	0.0182
41	0.0184	42	-45.0000	43	-44.8883	44	-44.7766
45	-44.4357	46	-44.0947	47	-43.5246	48	-42.9547
49	-42.1558	50	-41.3575	51	-40.3311	52	-39.3060
53	-38.0547	54	-36.8062	55	-35.3346	56	-33.8686
57	-32.1845	58	-30.5100	59	-28.6251	60	-26.7554
61	-24.6715	62	-22.6106	63	-20.3893	64	-18.1977
65	-15.8174	66	-13.4729	67	-10.9621	68	-8.4859
69	-5.8172	70	-0.1500	71	-0.1500	72	-0.1500
73	-0.1500	74	-0.1500	75	-0.1500	76	-0.1500
77	-0.1500	78	0.0166	79	0.0171	80	0.0177
81	0.0182	82	0.0184	83	-45.0000	84	-44.8883
85	-44.7766	86	-44.4357	87	-44.0947	88	-43.5246
89	-42.9547	90	-42.1558	91	-41.3575	92	-40.3311
93	-39.3060	94	-38.0547	95	-36.8062	96	-35.3346
97	-33.8686	98	-32.1845	99	-30.5100	100	-28.6251
101	-26.7554	102	-24.6715	103	-22.6106	104	-20.3893

```
105   -18.1977  106   -15.8174  107   -13.4729  108  -10.9621
109    -8.4859  110    -5.8172  111    -0.1500  112   -0.1500
113    -0.1500  114    -0.1500  115    -0.1500  116   -0.1500
117    -0.1500  118    -0.1500  119     0.0166  120    0.0171
121     0.0177  122     0.0182  123     0.0184
```

A.7.2   Steady State Forced Convection Example Using the SUPG Method

This example has been used to show the effectiveness of the SUPG method in Chapter 7. The problem details appear in Section 7.2.2 and Figure 7.3 of Chapter 7.

Input Data File

Contents of the **INPUTS.DAT** for this example are as follows:

```
Petrov-Galerkin benchmark
   0   0   0   0   1   1
   400  441   1   4
   1    2   23   22   1
   2    3   24   23   1
   3    4   25   24   1
   4    5   26   25   1
   5    6   27   26   1
   6    7   28   27   1
   7    8   29   28   1
   8    9   30   29   1
   9   10   31   30   1
  10   11   32   31   1
  11   12   33   32   1
  12   13   34   33   1
  13   14   35   34   1
  14   15   36   35   1
  15   16   37   36   1
  16   17   38   37   1
  17   18   39   38   1
  18   19   40   39   1
  19   20   41   40   1
  20   21   42   41   1
  22   23   44   43   1
  23   24   45   44   1
  24   25   46   45   1
  25   26   47   46   1
  26   27   48   47   1
  27   28   49   48   1
  28   29   50   49   1
  29   30   51   50   1
  30   31   52   51   1
  31   32   53   52   1
  32   33   54   53   1
  33   34   55   54   1
  34   35   56   55   1
  35   36   57   56   1
  36   37   58   57   1
  37   38   59   58   1
  38   39   60   59   1
  39   40   61   60   1
  40   41   62   61   1
  41   42   63   62   1
  43   44   65   64   1
```

```
44   45   66   65   1
45   46   67   66   1
46   47   68   67   1
47   48   69   68   1
48   49   70   69   1
49   50   71   70   1
50   51   72   71   1
51   52   73   72   1
52   53   74   73   1
53   54   75   74   1
54   55   76   75   1
55   56   77   76   1
56   57   78   77   1
57   58   79   78   1
58   59   80   79   1
59   60   81   80   1
60   61   82   81   1
61   62   83   82   1
62   63   84   83   1
64   65   86   85   1
65   66   87   86   1
66   67   88   87   1
67   68   89   88   1
68   69   90   89   1
69   70   91   90   1
70   71   92   91   1
71   72   93   92   1
72   73   94   93   1
73   74   95   94   1
74   75   96   95   1
75   76   97   96   1
76   77   98   97   1
77   78   99   98   1
78   79   100  99   1
79   80   101  100  1
80   81   102  101  1
81   82   103  102  1
82   83   104  103  1
83   84   105  104  1
85   86   107  106  1
86   87   108  107  1
87   88   109  108  1
88   89   110  109  1
89   90   111  110  1
90   91   112  111  1
91   92   113  112  1
92   93   114  113  1
93   94   115  114  1
94   95   116  115  1
95   96   117  116  1
96   97   118  117  1
97   98   119  118  1
98   99   120  119  1
99   100  121  120  1
100  101  122  121  1
101  102  123  122  1
102  103  124  123  1
103  104  125  124  1
104  105  126  125  1
106  107  128  127  1
107  108  129  128  1
108  109  130  129  1
109  110  131  130  1
110  111  132  131  1
111  112  133  132  1
112  113  134  133  1
113  114  135  134  1
```

```
114  115  136  135  1
115  116  137  136  1
116  117  138  137  1
117  118  139  138  1
118  119  140  139  1
119  120  141  140  1
120  121  142  141  1
121  122  143  142  1
122  123  144  143  1
123  124  145  144  1
124  125  146  145  1
125  126  147  146  1
127  128  149  148  1
128  129  150  149  1
129  130  151  150  1
130  131  152  151  1
131  132  153  152  1
132  133  154  153  1
133  134  155  154  1
134  135  156  155  1
135  136  157  156  1
136  137  158  157  1
137  138  159  158  1
138  139  160  159  1
139  140  161  160  1
140  141  162  161  1
141  142  163  162  1
142  143  164  163  1
143  144  165  164  1
144  145  166  165  1
145  146  167  166  1
146  147  168  167  1
148  149  170  169  1
149  150  171  170  1
150  151  172  171  1
151  152  173  172  1
152  153  174  173  1
153  154  175  174  1
154  155  176  175  1
155  156  177  176  1
156  157  178  177  1
157  158  179  178  1
158  159  180  179  1
159  160  181  180  1
160  161  182  181  1
161  162  183  182  1
162  163  184  183  1
163  164  185  184  1
164  165  186  185  1
165  166  187  186  1
166  167  188  187  1
167  168  189  188  1
169  170  191  190  1
170  171  192  191  1
171  172  193  192  1
172  173  194  193  1
173  174  195  194  1
174  175  196  195  1
175  176  197  196  1
176  177  198  197  1
177  178  199  198  1
178  179  200  199  1
179  180  201  200  1
180  181  202  201  1
181  182  203  202  1
182  183  204  203  1
183  184  205  204  1
```

```
184   185   206   205   1
185   186   207   206   1
186   187   208   207   1
187   188   209   208   1
188   189   210   209   1
190   191   212   211   1
191   192   213   212   1
192   193   214   213   1
193   194   215   214   1
194   195   216   215   1
195   196   217   216   1
196   197   218   217   1
197   198   219   218   1
198   199   220   219   1
199   200   221   220   1
200   201   222   221   1
201   202   223   222   1
202   203   224   223   1
203   204   225   224   1
204   205   226   225   1
205   206   227   226   1
206   207   228   227   1
207   208   229   228   1
208   209   230   229   1
209   210   231   230   1
211   212   233   232   1
212   213   234   233   1
213   214   235   234   1
214   215   236   235   1
215   216   237   236   1
216   217   238   237   1
217   218   239   238   1
218   219   240   239   1
219   220   241   240   1
220   221   242   241   1
221   222   243   242   1
222   223   244   243   1
223   224   245   244   1
224   225   246   245   1
225   226   247   246   1
226   227   248   247   1
227   228   249   248   1
228   229   250   249   1
229   230   251   250   1
230   231   252   251   1
232   233   254   253   1
233   234   255   254   1
234   235   256   255   1
235   236   257   256   1
236   237   258   257   1
237   238   259   258   1
238   239   260   259   1
239   240   261   260   1
240   241   262   261   1
241   242   263   262   1
242   243   264   263   1
243   244   265   264   1
244   245   266   265   1
245   246   267   266   1
246   247   268   267   1
247   248   269   268   1
248   249   270   269   1
249   250   271   270   1
250   251   272   271   1
251   252   273   272   1
253   254   275   274   1
254   255   276   275   1
```

255	256	277	276	1
256	257	278	277	1
257	258	279	278	1
258	259	280	279	1
259	260	281	280	1
260	261	282	281	1
261	262	283	282	1
262	263	284	283	1
263	264	285	284	1
264	265	286	285	1
265	266	287	286	1
266	267	288	287	1
267	268	289	288	1
268	269	290	289	1
269	270	291	290	1
270	271	292	291	1
271	272	293	292	1
272	273	294	293	1
274	275	296	295	1
275	276	297	296	1
276	277	298	297	1
277	278	299	298	1
278	279	300	299	1
279	280	301	300	1
280	281	302	301	1
281	282	303	302	1
282	283	304	303	1
283	284	305	304	1
284	285	306	305	1
285	286	307	306	1
286	287	308	307	1
287	288	309	308	1
288	289	310	309	1
289	290	311	310	1
290	291	312	311	1
291	292	313	312	1
292	293	314	313	1
293	294	315	314	1
295	296	317	316	1
296	297	318	317	1
297	298	319	318	1
298	299	320	319	1
299	300	321	320	1
300	301	322	321	1
301	302	323	322	1
302	303	324	323	1
303	304	325	324	1
304	305	326	325	1
305	306	327	326	1
306	307	328	327	1
307	308	329	328	1
308	309	330	329	1
309	310	331	330	1
310	311	332	331	1
311	312	333	332	1
312	313	334	333	1
313	314	335	334	1
314	315	336	335	1
316	317	338	337	1
317	318	339	338	1
318	319	340	339	1
319	320	341	340	1
320	321	342	341	1
321	322	343	342	1
322	323	344	343	1
323	324	345	344	1
324	325	346	345	1

325	326	347	346	1
326	327	348	347	1
327	328	349	348	1
328	329	350	349	1
329	330	351	350	1
330	331	352	351	1
331	332	353	352	1
332	333	354	353	1
333	334	355	354	1
334	335	356	355	1
335	336	357	356	1
337	338	359	358	1
338	339	360	359	1
339	340	361	360	1
340	341	362	361	1
341	342	363	362	1
342	343	364	363	1
343	344	365	364	1
344	345	366	365	1
345	346	367	366	1
346	347	368	367	1
347	348	369	368	1
348	349	370	369	1
349	350	371	370	1
350	351	372	371	1
351	352	373	372	1
352	353	374	373	1
353	354	375	374	1
354	355	376	375	1
355	356	377	376	1
356	357	378	377	1
358	359	380	379	1
359	360	381	380	1
360	361	382	381	1
361	362	383	382	1
362	363	384	383	1
363	364	385	384	1
364	365	386	385	1
365	366	387	386	1
366	367	388	387	1
367	368	389	388	1
368	369	390	389	1
369	370	391	390	1
370	371	392	391	1
371	372	393	392	1
372	373	394	393	1
373	374	395	394	1
374	375	396	395	1
375	376	397	396	1
376	377	398	397	1
377	378	399	398	1
379	380	401	400	1
380	381	402	401	1
381	382	403	402	1
382	383	404	403	1
383	384	405	404	1
384	385	406	405	1
385	386	407	406	1
386	387	408	407	1
387	388	409	408	1
388	389	410	409	1
389	390	411	410	1
390	391	412	411	1
391	392	413	412	1
392	393	414	413	1
393	394	415	414	1
394	395	416	415	1

```
395   396   417   416   1
396   397   418   417   1
397   398   419   418   1
398   399   420   419   1
400   401   422   421   1
401   402   423   422   1
402   403   424   423   1
403   404   425   424   1
404   405   426   425   1
405   406   427   426   1
406   407   428   427   1
407   408   429   428   1
408   409   430   429   1
409   410   431   430   1
410   411   432   431   1
411   412   433   432   1
412   413   434   433   1
413   414   435   434   1
414   415   436   435   1
415   416   437   436   1
416   417   438   437   1
417   418   439   438   1
418   419   440   439   1
419   420   441   440   1
0.    0.
   5.00000E-02   0.
   1.00000E-01   0.
   0.150000   0.
   0.200000   0.
   0.250000   0.
   0.300000   0.
   0.350000   0.
   0.400000   0.
   0.450000   0.
   0.500000   0.
   0.550000   0.
   0.600000   0.
   0.650000   0.
   0.700000   0.
   0.750000   0.
   0.800000   0.
   0.850000   0.
   0.900000   0.
   0.950000   0.
   1.00000    0.
0.    5.00000E-02
   5.00000E-02   5.00000E-02
   1.00000E-01   5.00000E-02
   0.150000   5.00000E-02
   0.200000   5.00000E-02
   0.250000   5.00000E-02
   0.300000   5.00000E-02
   0.350000   5.00000E-02
   0.400000   5.00000E-02
   0.450000   5.00000E-02
   0.500000   5.00000E-02
   0.550000   5.00000E-02
   0.600000   5.00000E-02
   0.650000   5.00000E-02
   0.700000   5.00000E-02
   0.750000   5.00000E-02
   0.800000   5.00000E-02
   0.850000   5.00000E-02
   0.900000   5.00000E-02
   0.950000   5.00000E-02
   1.00000    5.00000E-02
0.    1.00000E-01
```

```
     5.00000E-02     1.00000E-01
     1.00000E-01     1.00000E-01
     0.150000      1.00000E-01
     0.200000      1.00000E-01
     0.250000      1.00000E-01
     0.300000      1.00000E-01
     0.350000      1.00000E-01
     0.400000      1.00000E-01
     0.450000      1.00000E-01
     0.500000      1.00000E-01
     0.550000      1.00000E-01
     0.600000      1.00000E-01
     0.650000      1.00000E-01
     0.700000      1.00000E-01
     0.750000      1.00000E-01
     0.800000      1.00000E-01
     0.850000      1.00000E-01
     0.900000      1.00000E-01
     0.950000      1.00000E-01
     1.00000       1.00000E-01
  0.    0.150000
     5.00000E-02     0.150000
     1.00000E-01     0.150000
     0.150000      0.150000
     0.200000      0.150000
     0.250000      0.150000
     0.300000      0.150000
     0.350000      0.150000
     0.400000      0.150000
     0.450000      0.150000
     0.500000      0.150000
     0.550000      0.150000
     0.600000      0.150000
     0.650000      0.150000
     0.700000      0.150000
     0.750000      0.150000
     0.800000      0.150000
     0.850000      0.150000
     0.900000      0.150000
     0.950000      0.150000
     1.00000       0.150000
  0.    0.200000
     5.00000E-02     0.200000
     1.00000E-01     0.200000
     0.150000      0.200000
     0.200000      0.200000
     0.250000      0.200000
     0.300000      0.200000
     0.350000      0.200000
     0.400000      0.200000
     0.450000      0.200000
     0.500000      0.200000
     0.550000      0.200000
     0.600000      0.200000
     0.650000      0.200000
     0.700000      0.200000
     0.750000      0.200000
     0.800000      0.200000
     0.850000      0.200000
     0.900000      0.200000
     0.950000      0.200000
     1.00000       0.200000
  0.    0.250000
     5.00000E-02     0.250000
     1.00000E-01     0.250000
     0.150000      0.250000
     0.200000      0.250000
```

```
0.250000    0.250000
0.300000    0.250000
0.350000    0.250000
0.400000    0.250000
0.450000    0.250000
0.500000    0.250000
0.550000    0.250000
0.600000    0.250000
0.650000    0.250000
0.700000    0.250000
0.750000    0.250000
0.800000    0.250000
0.850000    0.250000
0.900000    0.250000
0.950000    0.250000
1.00000     0.250000
0.    0.300000
  5.00000E-02    0.300000
  1.00000E-01    0.300000
0.150000    0.300000
0.200000    0.300000
0.250000    0.300000
0.300000    0.300000
0.350000    0.300000
0.400000    0.300000
0.450000    0.300000
0.500000    0.300000
0.550000    0.300000
0.600000    0.300000
0.650000    0.300000
0.700000    0.300000
0.750000    0.300000
0.800000    0.300000
0.850000    0.300000
0.900000    0.300000
0.950000    0.300000
1.00000     0.300000
0.    0.350000
  5.00000E-02    0.350000
  1.00000E-01    0.350000
0.150000    0.350000
0.200000    0.350000
0.250000    0.350000
0.300000    0.350000
0.350000    0.350000
0.400000    0.350000
0.450000    0.350000
0.500000    0.350000
0.550000    0.350000
0.600000    0.350000
0.650000    0.350000
0.700000    0.350000
0.750000    0.350000
0.800000    0.350000
0.850000    0.350000
0.900000    0.350000
0.950000    0.350000
1.00000     0.350000
0.    0.400000
  5.00000E-02    0.400000
  1.00000E-01    0.400000
0.150000    0.400000
0.200000    0.400000
0.250000    0.400000
0.300000    0.400000
0.350000    0.400000
0.400000    0.400000
```

```
      0.450000   0.400000
      0.500000   0.400000
      0.550000   0.400000
      0.600000   0.400000
      0.650000   0.400000
      0.700000   0.400000
      0.750000   0.400000
      0.800000   0.400000
      0.850000   0.400000
      0.900000   0.400000
      0.950000   0.400000
      1.00000    0.400000
   0.   0.450000
      5.00000E-02   0.450000
      1.00000E-01   0.450000
      0.150000   0.450000
      0.200000   0.450000
      0.250000   0.450000
      0.300000   0.450000
      0.350000   0.450000
      0.400000   0.450000
      0.450000   0.450000
      0.500000   0.450000
      0.550000   0.450000
      0.600000   0.450000
      0.650000   0.450000
      0.700000   0.450000
      0.750000   0.450000
      0.800000   0.450000
      0.850000   0.450000
      0.900000   0.450000
      0.950000   0.450000
      1.00000    0.450000
   0.   0.500000
      5.00000E-02   0.500000
      1.00000E-01   0.500000
      0.150000   0.500000
      0.200000   0.500000
      0.250000   0.500000
      0.300000   0.500000
      0.350000   0.500000
      0.400000   0.500000
      0.450000   0.500000
      0.500000   0.500000
      0.550000   0.500000
      0.600000   0.500000
      0.650000   0.500000
      0.700000   0.500000
      0.750000   0.500000
      0.800000   0.500000
      0.850000   0.500000
      0.900000   0.500000
      0.950000   0.500000
      1.00000    0.500000
   0.   0.550000
      5.00000E-02   0.550000
      1.00000E-01   0.550000
      0.150000   0.550000
      0.200000   0.550000
      0.250000   0.550000
      0.300000   0.550000
      0.350000   0.550000
      0.400000   0.550000
      0.450000   0.550000
      0.500000   0.550000
      0.550000   0.550000
      0.600000   0.550000
```

```
  0.650000    0.550000
  0.700000    0.550000
  0.750000    0.550000
  0.800000    0.550000
  0.850000    0.550000
  0.900000    0.550000
  0.950000    0.550000
  1.00000     0.550000
0.    0.600000
  5.00000E-02   0.600000
  1.00000E-01   0.600000
  0.150000    0.600000
  0.200000    0.600000
  0.250000    0.600000
  0.300000    0.600000
  0.350000    0.600000
  0.400000    0.600000
  0.450000    0.000000
  0.500000    0.600000
  0.550000    0.600000
  0.600000    0.600000
  0.650000    0.600000
  0.700000    0.600000
  0.750000    0.600000
  0.800000    0.600000
  0.850000    0.600000
  0.900000    0.600000
  0.950000    0.600000
  1.00000     0.600000
0.    0.650000
  5.00000E-02   0.650000
  1.00000E-01   0.650000
  0.150000    0.650000
  0.200000    0.650000
  0.250000    0.650000
  0.300000    0.650000
  0.350000    0.650000
  0.400000    0.650000
  0.450000    0.650000
  0.500000    0.650000
  0.550000    0.650000
  0.600000    0.650000
  0.650000    0.650000
  0.700000    0.650000
  0.750000    0.650000
  0.800000    0.650000
  0.850000    0.650000
  0.900000    0.650000
  0.950000    0.650000
  1.00000     0.650000
0.    0.700000
  5.00000E-02   0.700000
  1.00000E-01   0.700000
  0.150000    0.700000
  0.200000    0.700000
  0.250000    0.700000
  0.300000    0.700000
  0.350000    0.700000
  0.400000    0.700000
  0.450000    0.700000
  0.500000    0.700000
  0.550000    0.700000
  0.600000    0.700000
  0.650000    0.700000
  0.700000    0.700000
  0.750000    0.700000
  0.800000    0.700000
```

```
        0.850000     0.700000
        0.900000     0.700000
        0.950000     0.700000
        1.00000      0.700000
  0.      0.750000
        5.00000E-02     0.750000
        1.00000E-01     0.750000
        0.150000     0.750000
        0.200000     0.750000
        0.250000     0.750000
        0.300000     0.750000
        0.350000     0.750000
        0.400000     0.750000
        0.450000     0.750000
        0.500000     0.750000
        0.550000     0.750000
        0.600000     0.750000
        0.650000     0.750000
        0.700000     0.750000
        0.750000     0.750000
        0.800000     0.750000
        0.850000     0.750000
        0.900000     0.750000
        0.950000     0.750000
        1.00000      0.750000
  0.      0.800000
        5.00000E-02     0.800000
        1.00000E-01     0.800000
        0.150000     0.800000
        0.200000     0.800000
        0.250000     0.800000
        0.300000     0.800000
        0.350000     0.800000
        0.400000     0.800000
        0.450000     0.800000
        0.500000     0.800000
        0.550000     0.800000
        0.600000     0.800000
        0.650000     0.800000
        0.700000     0.800000
        0.750000     0.800000
        0.800000     0.800000
        0.850000     0.800000
        0.900000     0.800000
        0.950000     0.800000
        1.00000      0.800000
  0.      0.850000
        5.00000E-02     0.850000
        1.00000E-01     0.850000
        0.150000     0.850000
        0.200000     0.850000
        0.250000     0.850000
        0.300000     0.850000
        0.350000     0.850000
        0.400000     0.850000
        0.450000     0.850000
        0.500000     0.850000
        0.550000     0.850000
        0.600000     0.850000
        0.650000     0.850000
        0.700000     0.850000
        0.750000     0.850000
        0.800000     0.850000
        0.850000     0.850000
        0.900000     0.850000
        0.950000     0.850000
        1.00000      0.850000
```

```
0.    0.900000
   5.00000E-02   0.900000
   1.00000E-01   0.900000
0.150000   0.900000
0.200000   0.900000
0.250000   0.900000
0.300000   0.900000
0.350000   0.900000
0.400000   0.900000
0.450000   0.900000
0.500000   0.900000
0.550000   0.900000
0.600000   0.900000
0.650000   0.900000
0.700000   0.900000
0.750000   0.900000
0.800000   0.900000
0.850000   0.900000
0.900000   0.900000
0.950000   0.900000
1.00000   0.900000
0.    0.950000
   5.00000E-02   0.950000
   1.00000E-01   0.950000
0.150000   0.950000
0.200000   0.950000
0.250000   0.950000
0.300000   0.950000
0.350000   0.950000
0.400000   0.950000
0.450000   0.950000
0.500000   0.950000
0.550000   0.950000
0.600000   0.950000
0.650000   0.950000
0.700000   0.950000
0.750000   0.950000
0.800000   0.950000
0.850000   0.950000
0.900000   0.950000
0.950000   0.950000
1.00000   0.950000
0.    1.00000
   5.00000E-02   1.00000
   1.00000E-01   1.00000
0.150000   1.00000
0.200000   1.00000
0.250000   1.00000
0.300000   1.00000
0.350000   1.00000
0.400000   1.00000
0.450000   1.00000
0.500000   1.00000
0.550000   1.00000
0.600000   1.00000
0.650000   1.00000
0.700000   1.00000
0.750000   1.00000
0.800000   1.00000
0.850000   1.00000
0.900000   1.00000
0.950000   1.00000
1.00000   1.00000
   1.0d-6   1.0
   41
   1    1.0
   2    1.0
```

```
  3    1.0
  4    1.0
  5    1.0
  6    1.0
  7    1.0
  8    1.0
  9    1.0
 10    1.0
 11    1.0
 12    1.0
 13    1.0
 14    1.0
 15    1.0
 16    1.0
 17    1.0
 18    1.0
 19    1.0
 20    1.0
 21    1.0
 22    1.0
 43    1.0
 64    1.0
 85    1.0
106    1.0
127    0.0
148    0.0
169    0.0
190    0.0
211    0.0
232    0.0
253    0.0
274    0.0
295    0.0
316    0.0
337    0.0
358    0.0
379    0.0
400    0.0
421    0.0
  0
  0
0.8660254  0.5
    .        .
    .        .
    .        .
0.8660254  0.5
```

The last set of data consists of *npoin* lines of velocities (u and v), which are all the same therefore only two lines are shown.

Output Data File

The output data file (**OUTPUT.RES**) for this example is listed below:

```
Petrov-Galerkin benchmark


    ITRAN =   0      ILINR =   0      IAXSY =   0

    ICONV =   1      IPETR =   1



   NELEM = 400      NPOIN = 441      NMATR =   1      NNPEL =   4
```

```
ELEMENT, MATERIAL,    NODE NUMBERS
   1       1         1   2   23   22
```

element information as in input data

```
  400       1        419 420 441 440
```

```
NODE      X         Y
  1     0.000     0.000
```

nodal coordinates as in input data

```
  441    1.000     1.000
```

```
MAT. CONDUCTIVITY CAPACITY
  1      0.000     1.000
```

NODE	FIXED TEMPR
1	1.0000
2	1.0000
3	1.0000
4	1.0000
5	1.0000
6	1.0000
7	1.0000
8	1.0000
9	1.0000
10	1.0000
11	1.0000
12	1.0000
13	1.0000
14	1.0000
15	1.0000
16	1.0000
17	1.0000
18	1.0000
19	1.0000
20	1.0000
21	1.0000
22	1.0000
43	1.0000
64	1.0000
85	1.0000
106	1.0000
127	0.0000
148	0.0000
169	0.0000
190	0.0000
211	0.0000
232	0.0000
253	0.0000
274	0.0000
295	0.0000
316	0.0000
337	0.0000
358	0.0000
379	0.0000
400	0.0000
421	0.0000

SPECIFIED VELOCITY FIELD

NODE	U-VEL	V-VEL
1	0.8660	0.5000
.	.	.
.	.	.
.	.	.

441   0.8660   0.5000

SOLUTION CONVERGED AT ITERATION NO.   1

*** OUTPUT OF TEMPERATURE ***

1	1.0000	2	1.0000	3	1.0000	4	1.0000
5	1.0000	6	1.0000	7	1.0000	8	1.0000
9	1.0000	10	1.0000	11	1.0000	12	1.0000
13	1.0000	14	1.0000	15	1.0000	16	1.0000
17	1.0000	18	1.0000	19	1.0000	20	1.0000
21	1.0000	22	1.0000	23	1.0000	24	0.9999
25	0.9997	26	0.9997	27	0.9999	28	1.0001
29	1.0003	30	1.0003	31	1.0003	32	1.0002
33	1.0001	34	1.0000	35	1.0000	36	0.9999
37	1.0000	38	1.0000	39	1.0000	40	1.0000
41	1.0000	42	1.0000	43	1.0000	44	1.0005
45	1.0008	46	1.0005	47	0.9997	48	0.9988
49	0.9984	50	0.9986	51	0.9991	52	0.9997
53	1.0002	54	1.0004	55	1.0004	56	1.0002
57	1.0001	58	1.0000	59	0.9999	60	0.9999
61	1.0000	62	1.0000	63	1.0000	64	1.0000
65	0.9985	66	1.0005	67	1.0039	68	1.0062
69	1.0061	70	1.0042	71	1.0016	72	0.9996
73	0.9986	74	0.9987	75	0.9992	76	0.9998
77	1.0003	78	1.0004	79	1.0003	80	1.0002
81	1.0000	82	0.9999	83	0.9999	84	0.9999
85	1.0000	86	0.9915	87	0.9778	88	0.9736
89	0.9796	90	0.9901	91	0.9994	92	1.0044
93	1.0052	94	1.0033	95	1.0009	96	0.9993
97	0.9987	98	0.9989	99	0.9996	100	1.0001
101	1.0004	102	1.0004	103	1.0002	104	1.0001
105	0.9999	106	1.0000	107	1.1024	108	1.1031
109	1.0593	110	1.0135	111	0.9861	112	0.9792
113	0.9850	114	0.9946	115	1.0019	116	1.0047
117	1.0041	118	1.0019	119	0.9999	120	0.9989
121	0.9989	122	0.9994	123	0.9999	124	1.0003
125	1.0004	126	1.0003	127	0.0000	128	0.5055
129	0.8586	130	1.0405	131	1.0906	132	1.0689
133	1.0270	134	0.9949	135	0.9817	136	0.9836
137	0.9919	138	0.9998	139	1.0039	140	1.0042
141	1.0025	142	1.0005	143	0.9992	144	0.9989
145	0.9992	146	0.9998	147	1.0002	148	0.0000
149	0.0001	150	0.2605	151	0.5950	152	0.8700
153	1.0300	154	1.0831	155	1.0694	156	1.0319
157	0.9997	158	0.9841	159	0.9835	160	0.9905
161	0.9984	162	1.0031	163	1.0041	164	1.0028
165	1.0009	166	0.9994	167	0.9989	168	0.9992
169	0.0000	170	-0.0219	171	-0.0258	172	0.1138
173	0.3729	174	0.6610	175	0.8943	176	1.0322
177	1.0794	178	1.0676	179	1.0330	180	1.0020
181	0.9857	182	0.9838	183	0.9899	184	0.9974
185	1.0024	186	1.0039	187	1.0030	188	1.0011
189	0.9996	190	0.0000	191	-0.0011	192	-0.0239
193	-0.0410	194	0.0252	195	0.2035	196	0.4571
197	0.7165	198	0.9198	199	1.0381	200	1.0774
201	1.0650	202	1.0323	203	1.0028	204	0.9868
205	0.9842	206	0.9897	207	0.9969	208	1.0020
209	1.0038	210	1.0031	211	0.0000	212	0.0019
213	0.0007	214	-0.0185	215	-0.0431	216	-0.0230
217	0.0849	218	0.2822	219	0.5285	220	0.7650
221	0.9439	222	1.0450	223	1.0760	224	1.0619
225	1.0306	226	1.0026	227	0.9873	228	0.9847
229	0.9897	230	0.9966	231	1.0018	232	0.0000
233	0.0002	234	0.0027	235	0.0030	236	-0.0109
237	-0.0365	238	-0.0433	239	0.0101	240	0.1466

241	0.3542	242	0.5915	243	0.8080	244	0.9659
245	1.0516	246	1.0747	247	1.0586	248	1.0283
249	1.0019	250	0.9876	251	0.9851	252	0.9899
253	0.0000	254	-0.0002	255	0.0001	256	0.0027
257	0.0045	258	-0.0039	259	-0.0260	260	-0.0458
261	-0.0303	262	0.0514	263	0.2085	264	0.4210
265	0.6482	266	0.8463	267	0.9855	268	1.0573
269	1.0732	270	1.0550	271	1.0256	272	1.0008
273	0.9874	274	0.0000	275	0.0000	276	-0.0003
277	-0.0001	278	0.0022	279	0.0050	280	0.0013
281	-0.0152	282	-0.0385	283	-0.0460	284	-0.0077
285	0.0979	286	0.2699	287	0.4834	288	0.6996
289	0.8805	290	1.0026	291	1.0620	292	1.0712
293	1.0514	294	1.0229	295	0.0000	296	0.0000
297	0.0000	298	-0.0004	299	-0.0004	300	0.0014
301	0.0046	302	0.0043	303	-0.0061	304	-0.0272
305	-0.0460	306	-0.0381	307	0.0222	308	0.1476
309	0.3301	310	0.5419	311	0.7464	312	0.9109
313	1.0173	314	1.0654	315	1.0691	316	0.0000
317	0.0000	318	0.0000	319	0.0000	320	-0.0004
321	-0.0006	322	0.0006	323	0.0036	324	0.0055
325	0.0004	326	-0.0157	327	-0.0375	328	-0.0479
329	-0.0229	330	0.0577	331	0.1995	332	0.3888
333	0.5967	334	0.7888	335	0.9371	336	1.0283
337	0.0000	338	0.0000	339	0.0000	340	0.0001
341	0.0000	342	-0.0003	343	-0.0007	344	-0.0001
345	0.0024	346	0.0053	347	0.0042	348	-0.0062
349	-0.0258	350	-0.0448	351	-0.0440	352	-0.0015
353	0.0975	354	0.2525	355	0.4458	356	0.6472
357	0.8235	358	0.0000	359	0.0000	360	0.0000
361	0.0000	362	0.0001	363	0.0001	364	-0.0002
365	-0.0006	366	-0.0005	367	0.0012	368	0.0043
369	0.0057	370	0.0005	371	-0.0144	372	-0.0350
373	-0.0483	374	-0.0345	375	0.0254	376	0.1409
377	0.3061	378	0.4970	379	0.0000	380	0.0000
381	0.0000	382	0.0000	383	0.0000	384	0.0001
385	0.0001	386	-0.0001	387	-0.0005	388	-0.0007
389	0.0003	390	0.0029	391	0.0056	392	0.0043
393	-0.0052	394	-0.0232	395	-0.0424	396	-0.0475
397	-0.0192	398	0.0576	399	0.1865	400	0.0000
401	0.0000	402	0.0000	403	0.0000	404	0.0000
405	0.0000	406	0.0000	407	0.0001	408	0.0000
409	-0.0004	410	-0.0008	411	-0.0004	412	0.0016
413	0.0046	414	0.0058	415	0.0010	416	-0.0124
417	-0.0319	418	-0.0472	419	-0.0404	420	0.0043
421	0.0000	422	0.0000	423	0.0000	424	0.0000
425	0.0000	426	0.0000	427	0.0000	428	0.0000
429	0.0001	430	0.0001	431	-0.0002	432	-0.0007
433	-0.0008	434	0.0006	435	0.0035	436	0.0060
437	0.0044	438	-0.0049	439	-0.0224	440	-0.0419
441	-0.0423						

References

[1] C.Taylor and T.G.Hughes. *Finite Element Programming of the Navier-Stokes Equations.* Pineridge Press, Swansea, U.K., 1981.

[2] W.Weaver Jr. and P.R.Johnston. *Finite Elements for Structural Analysis.* Prentice-Hall, Englewood Cliffs, 1984.

[3] H.S.Carslaw and J.C.Jaeger. *Conduction of Heat in Solids.* Clarendon Press, Oxford, 1959.

Appendix B

Software Description for HADAPT

B.1 Introduction

In this appendix a detailed description of the program **HADAPT** is presented with complete instructions for the user. This program allows a user to perform *adaptive* heat transfer analysis for 2-D plane or axisymmetric problems. The program uses three types of elements, *i.e.*, 3 and 4-noded linear and 6-noded quadratic elements.

B.2 Glossary of Variable Names

A brief description of the main variables and arrays and the main subroutines is listed in the following sections.

B.2.1 Main Variables for the Geometry and Mesh Data

Variable name	Description
nbnds	Number of internal regions (holes)
natrib	Number of subdomains
toler	A tolerance value
nnpe	Number of nodes per element (3, 4 or 6)
ifrnt	Choice of optimisation (0-Profile optimisation, 1-Front optimization)
noseg	Number of boundary segments in current subdomain
nbpln	Total number of boundary segments
iatt(natrib)	Attribute number for each subdomain
ityp	Type of current boundary segment

Variable name	Description
x1,y1,x2,y2	Coordinate of end points of a line or an arc segment (ityp=1,2)
rad	Radius of an arc segment (for ityp=2) (-ve is clockwise)
numpt	Number of points in a user defined segment (ityp=3)
xx(numpt), yy(numpt)	Coordinate of each point for a user defined segment (for ityp=3)
iseg	Number of a previously defined segment (ityp=4)
ndpt	Number of mesh density specification points
xd(ndpt), yd(ndpt), vald(ndpt)	Coordinates of each density specification point and the corresponding mesh density value (element size) of each density specification point

B.2.2   Additional Variables Used in This Program

Variable name	Description
amult	Multiplies all required element sizes by 2.0 if $nnpel = 4$
elmax	Maximum permitted element size
elmin	Miniimum permitted element size
ersrt	The calculated average norm-error for the mesh
iadap	0 for the first call of the mesh generator, 1 for subsequent calls
ierst	1 if error estimation required, 0 otherwise
mbpln	Maximum number of boundary segments for a geometry
mele3	Maximum size of auxiliary array used in subroutine TRANSF
mpnod	Maximum number of nodes or element on a boundary segment
nobcd	Number of different fixed temperatures for Dirichlet BC's
nobcn	Number of different sets of quantities for Neumann BC's
num	Number of boundary segments on which BC's are to be applied
numb	Boundary segment number on which a BC is to be applied
numc	The corresponding BC number (of the *nobcd* or *nobcn* defined)
pcent	The target error to be reached by adapting meshes
rcent	A modified value of *pcent* actually satisfied

B.2.3   Additional Arrays Used in This Program

Variable name	Description
ambip(mcdpt)	Values of ambient temperature to be used for Neumann BC's (*nobcn* number)
coefp(mcdpt)	Values of convective heat transfer coefficient to be used for Neumann BC's (*nobcn* number)
coor0(mpoin,mdime)	Nodal coordinates for the previous mesh
darep(mele3)	Auxilliary array used in subroutine TRANSF
ehnew(mpoin)	Predicted element sizes around nodal points
ehold(melem)	Previous element sizes
emass(mnpel,mnpel)	Mass matrix of elements
erelm(melem)	Relative norm errors in each element
ernor(melem)	Square of norm errors in each element
fluxp(mcdpt)	Values of heat flux to be used for Neumann BC's (*nobcn* number)
icont(mpoin)	Auxilliary array used in subroutine ERREST
itemp(mpoin)	Auxilliary array used in subroutine TOMESH
lnod3(mele3)	Auxilliary array used in subroutine TRANSF
nbeln(mbpln)	Number of elements on each boundary segment with one face (edge) on an external boundary
nbctm(mbpln)	Array specifying the Dirichlet BC's on each boundary segment
nbctm(mbpln)	Array specifying the Neumann BC's on each boundary segment
nbele(mbpln,mpnod)	Element number of each element on each boundary segment
nbelt(mbpln,mpnod)	Auxilliary array for *nbele*
nbelf(mbpln,mpnod)	Face number of the external face of each element on each boundary segment
nbpoi(mbpln)	Number of nodes on each boundary segment
nbref(mbpln,mpnod)	Node number of each node on each boundary segment
nbret(mbpln,mpnod)	Auxilliary array for *nbref*
ncon0(melem,mnpel)	Element nodal connectivities for the previous mesh
newel(melem)	Array used in front optimization after mesh generation
newno(mpoin)	Array used in profile optimization after mesh generation
permi(melem)	Calculated element sizes (*nobcn* number)
radip(mcdpt)	Values of radiation coefficient to be used for Neumann BC's
sgrdx(melem,mgaus)	Smoothed values of dT/dx at Gaussian points
sgrdy(melem,mgaus)	Smoothed values of dT/dy at Gaussian points

Variable name	Description
smoth(mpoin)	Smoothed values of temperature gradients at nodal points
tener(melem)	Heat flow dissipation in each element
tgrdx(melem,mgaus)	Temperature gradients (dT/dx) at Gaussian points
tgrdy(melem,mgaus)	Temperature gradients (dT/dy) at Gaussian points
tembc(mcdpt)	Values of fixed temperatures to be applied as Dirichlet BC's (*nobcd* number)
temin(mmatr)	The temperature value to be used as initial condition for all the nodes in each material
tempa(mpoin)	Auxilliary array used in subroutine TRANSF
tempb(mpoin)	Auxilliary array used in subroutine TRANSF
tempc(mpoin)	Auxilliary array used in subroutine TRANSF
tempd(mpoin)	Auxilliary array used in subroutine TRANSF
uvelm(mmatr)	The X-component of velocity to be used for convection for all the nodes in each material
vvelm(mmatr)	The Y-component of velocity to be used for convection for all the nodes in each material
xdens(mpoin)	X-coordinates of all nodes transferred to the mesh generator for locating density points
ydens(mpoin)	Y-coordinates of all nodes transferred to the mesh generator for locating density points

B.2.4   New Subroutines Used in This Program

Subroutine	Description
CONTRL	Reads control data for the problem
MESH2D	Reads the geometry data, generates and optimizes the mesh
INDATA	Reads the problem data regarding material, boundary, and initial conditions
TOMESH	Transfers all problem data to the appropriate nodes and elements of the mesh
ERREST	Performs error estimation calculations
TRANSF	Transfers all necessary information from the old mesh to the new mesh

B.3   Program Overview

The main routine is listed in the following lines, the program structure
with all the subroutines is shown in Figure B.1.

```
      PROGRAM HADAPT                                              ADPT   1
C_____ADPT   2
C THIS PROGRAM PERFORMS H-ADAPTIVE 2-D FEM HEAT TRANSFER ANALYSIS. IT  ADPT  3
C INCLUDES AN AUTOMATIC MESH GENERATOR, MEHS2D FOR GENERATING TRIANGULARADPT  4
C (3 & 6 NODED ELEMENTS) AND QUADRILATERAL (4 NODED) ELEMENTS.    ADPT   5
C                                     A S USMANI AND H C HUANG  OCT 1993 ADPT  6
C_____ADPT   7
C                                                                ADPT   8
      PARAMETER (MELEM=4999,MPOIN=4999,MNPEL=9,MMATR=2,MBOUN=999,  ADPT   9
     -           MCDPT=10,MDIME=2,MGAUS=9,MPROF=199999,MDENS=MPOIN, ADPT  10
     -           MFRON=199,MBPLN=49,MPNOD=199,MELE3=MELEM*4)       ADPT  11
      IMPLICIT REAL*8(A-H,O-Z)                                    ADPT  12
      DIMENSION MTYPE(MELEM),NCONC(MELEM,MNPEL),COORD(MPOIN,MDIME), ADPT  13
     -          CONDY(MMATR),CAPCY(MMATR),FIXED(MBOUN),NFIXD(MBOUN), ADPT  14
     -          NFACB(MBOUN),NELMB(MBOUN),FLUXE(MBOUN),            ADPT  15
     -          COEFF(MBOUN),RADIA(MBOUN),AMBIT(MBOUN),            ADPT  16
     -          TEMPR(MPOIN),CDVLU(MCDPT,MMATR),CPVLU(MCDPT,MMATR), ADPT  17
     -          TVALU(MCDPT,MMATR)                                 ADPT  18
      DIMENSION SHAPF(MNPEL,MGAUS,MELEM),DERV1(MNPEL,MGAUS,MELEM), ADPT  19
     -          DERV2(MNPEL,MGAUS,MELEM),ARWET(MGAUS,MELEM),       ADPT  20
     -          WEIGB(MGAUS),POSGB(MGAUS),WEIGP(MGAUS),POSGX(MGAUS), ADPT  21
     -          POSGY(MGAUS)                                       ADPT  22
      DIMENSION NDFEL(MNPEL),ASTIF(MNPEL,MNPEL),AMASS(MNPEL,MNPEL), ADPT  23
     -          NCOLM(MPOIN),IFFIX(MPOIN),TFIXD(MPOIN),TEMPT(MNPEL), ADPT  24
     -          TEMP1(MNPEL,MNPEL),RVECT(MNPEL),NBELM(MELEM,4),    ADPT  25
     -          XCORD(MNPEL),YCORD(MNPEL),EFORC(MNPEL)             ADPT  26
      DIMENSION GSTIF(MPROF),NDIAG(MPOIN),FORCE(MPOIN),TLAST(MPOIN) ADPT  27
      DIMENSION TENER(MELEM),TGRDX(MELEM,MGAUS),TGRDY(MELEM,MGAUS), ADPT  28
     .          ERHS1(MELEM,MNPEL),ERHS2(MELEM,MNPEL),NDIAG(MPOIN), ADPT  29
     .          ERNOR(MELEM),SGRDX(MELEM,MGAUS),SGRDY(MELEM,MGAUS), ADPT  30
     .          EBELM(MELEM),SMOTH(MPOIN),PERMI(MELEM),EHOLD(MELEM), ADPT  31
     .          EHNEW(MDENS),XDENS(MDENS),YDENS(MDENS),ICONT(MPOIN) ADPT  32
      DIMENSION UVELO(MPOIN),VVELO(MPOIN),PETRV(MELEM),ULOCA(MNPEL), ADPT  33
     -          VLOCA(MNPEL),NDEST(MNPEL),LOCEL(MNPEL),LHEDV(MFRON), ADPT  34
     -          NADFM(MPOIN),NODFM(MPOIN),LNODS(MELEM,MNPEL),      ADPT  35
     -          PNORM(MFRON),GFLUM(MFRON,MFRON)                    ADPT  36
      DIMENSION NBCTM(MBPLN),NBCFL(MBPLN),TEMBC(MCDPT),FLUXP(MCDPT), ADPT  37
     -          COEFP(MCDPT),RADIP(MCDPT),AMBIP(MCDPT),TEMIN(MMATR), ADPT  38
     -          UVELM(MMATR),VVELM(MMATR)                          ADPT  39
      DIMENSION NBPOI(MBPLN),NBREF(MBPLN,MPNOD),NBELN(MBPLN),      ADPT  40
     .          NBELF(MBPLN,MPNOD),NBELE(MBPLN,MPNOD),ITEMP(MPOIN), ADPT  41
     .          NBRET(MBPLN,MPNOD),NBELT(MBPLN,MPNOD),NEWEL(MELEM), ADPT  42
     .          NEWNO(MPOIN)                                       ADPT  43
      DIMENSION LNOD3(MELE3,3),NCONO(MELEM,MNPEL),COORO(MPOIN,MDIME), ADPT  44
     .          DAREP(MELE3),TEMPA(MPOIN)                          ADPT  45
C                                                                ADPT  46
C *** OPEN CHANNELS FOR READING AND WRITING                       ADPT  47
C                                                                ADPT  48
      OPEN (UNIT=12,STATUS='OLD',FILE='ADPINP.DAT')               ADPT  49
      OPEN (UNIT=7,STATUS='NEW',FILE='ADPOUT.RES')               ADPT  50
C                                                                ADPT  51
      OPEN(UNIT=11,STATUS='OLD',FILE='GEOMET.DAT')               ADPT  52
      OPEN(UNIT=14,STATUS='OLD',FILE='DENSIT.DAT')               ADPT  53
      OPEN(UNIT=13,STATUS='UNKNOWN',FILE='GENERR.LOG')           ADPT  54
C                                                                ADPT  55
C *** READ CONTROL DATA                                          ADPT  56
C                                                                ADPT  57
      NSTEP=0                                                     ADPT  58
      IFRNT=0                                                     ADPT  59
```

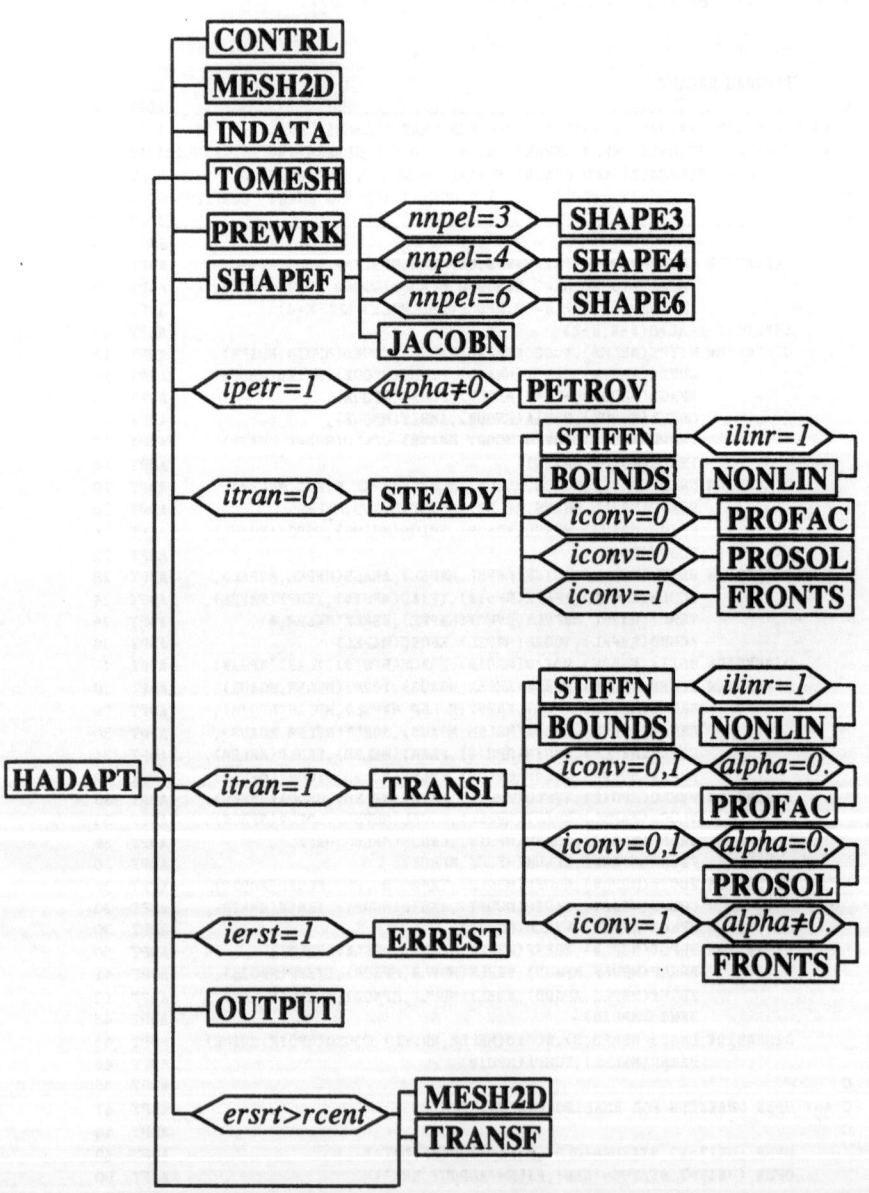

Figure B.1: HADAPT Program structure

```
        ITRAD=0                                                         ADPT  60
        NITRA=19                                                        ADPT  61
        ICOAR=1                                                         ADPT  62
        PCENT=1.0                                                       ADPT  63
        CALL CONTRL      (ITRAN,ILINR,IAXSY,IERST,ICONV,IPETR,NITER,    ADPT  64
      -                  NITUP,NITDN,TTIME,STIME,DTIME,DTMAX,ALPHA,      ADPT  65
      -                  RELAX,TOLER,PCENT,ELMIN,ELMAX,FACTR)           ADPT  66
        TIME=STIME                                                      ADPT  67
        RCENT=PCENT                                                     ADPT  68
        IF (ICONV.EQ.1.AND.ITRAN.EQ.0) IFRNT=1                          ADPT  69
        IF ((ICONV.EQ.1.AND.ITRAN.EQ.1).AND.ALPHA.GT.0.0DO) IFRNT=1     ADPT  70
        IF (IFRNT.EQ.1.AND.IERST.EQ.1) IFRNT=2                          ADPT  71
C                                                                       ADPT  72
        IF (IFRNT.EQ.1) THEN                                            ADPT  73
          OPEN (UNIT=2,STATUS='SCRATCH',FORM='UNFORMATTED')             ADPT  74
          OPEN (UNIT=4,STATUS='SCRATCH',FORM='UNFORMATTED')             ADPT  75
        END IF                                                          ADPT  76
C                                                                       ADPT  77
C *** READ IN GEOMETRY DATA AND CREATE MESH                             ADPT  78
C                                                                       ADPT  79
        IADAP=0                                                         ADPT  80
        CALL MESH2D      (MPOIN,MELEM,MNPEL,MBPLN,MPNOD,MDIME,MDENS,     ADPT  81
      -                  IADAP,IFRNT,NPOIN,NELEM,NNPEL,NMATR,NDENS,      ADPT  82
      -                  NBPLN,NBPOI,NBREF,NBELN,NBELF,NBELE,NCONC,      ADPT  83
      -                  MTYPE,NBELT,NBRET,NEWEL,NEWNO,COORD,XDENS,      ADPT  84
      -                  YDENS,EHNEW)                                    ADPT  85
        ANULT=1.0DO                                                     ADPT  86
        IF (NNPEL.EQ.4) ANULT=2.0DO                                     ADPT  87
C                                                                       ADPT  88
C *** READ PROBLEM DATA                                                 ADPT  89
C                                                                       ADPT  90
        CALL INDATA      (MMATR,MBPLN,MCDPT,NMATR,NCDPT,NOBCD,NOBCN,     ADPT  91
      -                  NBPLN,ITRAN,ILINR,ICONV,                       ADPT  92
      -                  NBCTN,NBCFL,TENBC,FLUXP,                       ADPT  93
      -                  COEFP,RADIP,AMBIP,TEMIN,CONDY,CAPCY,CDVLU,      ADPT  94
      -                  CPVLU,TVALU,UVELN,VVELN)                       ADPT  95
C                                                                       ADPT  96
C *** TRANSFER PROBLEM DATA TO MESH                                     ADPT  97
C                                                                       ADPT  98
   888  CALL TOMESH      (MPOIN,MELEM,MNPEL,MMATR,MBPLN,MCDPT,MPNOD,     ADPT  99
      -                  MDIME,MBOUN,ITRAN,ILINR,ICONV,NPOIN,NELEM,      ADPT 100
      -                  NNPEL,NMATR,NOBCD,NOBCN,NFIXB,NEUMN,NBPLN,      ADPT 101
      -                  NCONC,MTYPE,NBPOI,NBREF,NBELN,NBELF,NBELE,      ADPT 102
      -                  NBCTN,NBCFL,NFIXD,NELMB,NFACB,FIXED,TEMBC,      ADPT 103
      -                  FLUXE,COEFF,RADIA,AMBIT,FLUXP,COEFP,RADIP,      ADPT 104
      -                  AMBIP,TEMPR,TEMIN,UVELO,VVELO,UVELN,VVELN,      ADPT 105
      -                  COORD,ITEMP)                                   ADPT 106
C                                                                       ADPT 107
C *** CREATE REQUIRED INFORMATION                                       ADPT 108
C                                                                       ADPT 109
        CALL PREWRK      (MELEM,MPOIN,MNPEL,MBOUN,NELEM,NPOIN,NNPEL,     ADPT 110
      -                  ICONV,ITRAN,NPROF,NCONC,NDIAG,NCOLM,NELMB,      ADPT 111
      -                  NBELM,IFFIX,NFIXD,NEUMN,NNPFC,NFIXB,LNODS,      ADPT 112
      -                  NADFM,NODFM,TFIXD,FIXED,ALPHA)                 ADPT 113
C                                                                       ADPT 114
C *** IF SUPG METHOD IS TO BE USED FOR CONVECTION CALCULATE THE SCALAR  ADPT 115
C *** COEFFICIENT FOR EACH ELEMENT                                      ADPT 116
C                                                                       ADPT 117
        IF (ICONV.EQ.1.AND.IPETR.EQ.1) THEN                            ADPT 118
        IF (ITRAN.EQ.1.AND.ALPHA.EQ.0.0) GOTO 311                      ADPT 119
        CALL PETROV      (MPOIN,MELEM,MNPEL,MDIME,MMATR,NPOIN,NELEM,     ADPT 120
      -                  NNPEL,NNPFC,NCONC,MTYPE,CONDY,COORD,PETRV,      ADPT 121
      -                  UVELO,VVELO)                                   ADPT 122
   311  CONTINUE                                                        ADPT 123
        END IF                                                          ADPT 124
C                                                                       ADPT 125
C *** CALCULATE ELEMENT SHAPE FUNCTIONS AND DERIVATIVES                 ADPT 126
```

```
C                                                                   ADPT 127
      CALL SHAPEF        (NELEM,NPOIN,NDIME,NNPEL,NGAUS,NELEM,NPOIN,  ADPT 128
     -                   NNPEL,NGASB,NGAUS,NCONC,COORD,WEIGB,POSGB,   ADPT 129
     -                   WEIGP,POSGX,POSGY,SHAPF,DERV1,DERV2,ARWET,   ADPT 130
     -                   IAXSY)                                       ADPT 131
      IF (ITRAN.EQ.1) GOTO 100                                       ADPT 132
C                                                                   ADPT 133
C *** PERFORM  STEADY STATE  ANALYSIS                                ADPT 134
C                                                                   ADPT 135
      CALL STEADY        (NELEM,NPOIN,NGAUS,NNPEL,NCDPT,NMATR,NTYPE,  ADPT 136
     -                   NDIME,NBOUN,NPROF,NFRON,IAXSY,ILINR,ITRAN,   ADPT 137
     -                   ICONV,IPETR,NITER,NFIXB,TOLER,RELAX,         ADPT 138
     -                   NELEM,NPOIN,NGASB,NGAUS,NNPEL,               ADPT 139
     -                   NNPFC,NPROF,NCDPT,NDFEL,NFACB,NBELM,NCONC,   ADPT 140
     -                   LNODS,NDIAG,IFFIX,TFIXD,TVALU,CDVLU,CPVLU,   ADPT 141
     -                   CONDY,CAPCY,TEMPT,SHAPF,DERV1,DERV2,ARWET,   ADPT 142
     -                   ASTIF,AMASS,POSGB,WEIGB,GSTIF,XCORD,YCORD,   ADPT 143
     -                   EFORC,FORCE,FLUXE,COEFF,RADIA,AMBIT,TLAST,   ADPT 144
     -                   TEMPR,COORD,UVELO,VVELO,ULOCA,VLOCA,PETRV,   ADPT 145
     -                   NDEST,LOCEL,LHEDV,NADFN,NODFN,PNORM,GFLUM)   ADPT 146
C                                                                   ADPT 147
C*** CONDUCT ERROR ESTIMATION IF REQUIRED                            ADPT 148
C                                                                   ADPT 149
      IF(IERST.NE.0)                                                 ADPT 150
     -CALL ERREST        (NELEM,NGAUS,NNPEL,NCDPT,NMATR,NPOIN,NGAUS,  ADPT 151
     -                   NNPEL,ILINR,ICOAR,NCDPT,TVALU,CDVLU,SHAPF,   ADPT 152
     -                   DERV1,DERV2,NPOIN,NPROF,ARWET,NDIME,COORD,   ADPT 153
     -                   NTYPE,CONDY,TEMPR,NCONC,NELEM,GSTIF,IFFIX,   ADPT 154
     -                   FORCE,TENER,TGRDX,TGRDY,ERHS1,ERHS2,ERNOR,   ADPT 155
     -                   ERELN,SGRDX,SGRDY,SMOTH,NDIAG,NCOLN,PERMI,   ADPT 156
     -                   EHOLD,EHNEW,ICONT,ERDOM,ERSRT,ERMAX,KELRM,   ADPT 157
     -                   PCENT)                                       ADPT 158
C                                                                   ADPT 159
C *** OUTPUT RESULTS                                                 ADPT 160
C                                                                   ADPT 161
      CALL  OUTPUT       (NELEM,NPOIN,NMATR,NNPEL,NPOIN,NELEM,NNPEL,  ADPT 162
     .                   NMATR,IERST,NCONC,NTYPE,TEMPR,COORD,NDIME,   ADPT 163
     .                   ERELN,ERDOM,ERSRT,EHNEW,ERMAX,KELRM,UVELO,   ADPT 164
     .                   VVELO,TIME)                                  ADPT 165
C                                                                   ADPT 166
C *** IF REQUIRED ERROR NOT ATTAINED REMESH AND REANALYSE            ADPT 167
C                                                                   ADPT 168
      IF (ERSRT.LE.RCENT) GOTO 999                                   ADPT 169
      ITRAD=ITRAD+1                                                  ADPT 170
      IF (ITRAD.GT.NITRA) WRITE(7,*)'MAX PERMITTED ADAPTIVE CYCLES OVER'ADPT 171
      IF (ITRAD.GT.NITRA) GOTO 999                                   ADPT 172
      DO I=1,NPOIN                                                   ADPT 173
         EHNEW(I)=EHNEW(I)*AMULT                                     ADPT 174
         IF (EHNEW(I).GT.ELMAX) EHNEW(I)=ELMAX                       ADPT 175
         IF (EHNEW(I).LT.ELMIN) EHNEW(I)=ELMIN                       ADPT 176
      END DO                                                        ADPT 177
      DO I=1,NPOIN                                                   ADPT 178
         XDENS(I)=COORD(I,1)                                         ADPT 179
         YDENS(I)=COORD(I,2)                                         ADPT 180
      END DO                                                        ADPT 181
      NDENS=NPOIN                                                    ADPT 182
C                                                                   ADPT 183
C *** TRANSFER OLD MESH INFO TO SEPERATE ARRAYS                      ADPT 184
C                                                                   ADPT 185
      NELEO=NELEM                                                    ADPT 186
      NPOIO=NPOIN                                                    ADPT 187
      DO I=1,NELEM                                                   ADPT 188
         DO J=1,NNPEL                                                ADPT 189
            NCONO(I,J)=NCONC(I,J)                                    ADPT 190
         END DO                                                     ADPT 191
      END DO                                                        ADPT 192
      DO I=1,NPOIN                                                   ADPT 193
```

```
              COORO(I,1)=COORD(I,1)                                    ADPT 194
              COORO(I,2)=COORD(I,2)                                    ADPT 195
          END DO                                                       ADPT 196
C                                                                      ADPT 197
C *** TRANSFER NODAL VALUES TO NEW MESH                                ADPT 198
C                                                                      ADPT 199
      REWIND 11                                                        ADPT 200
      IADAP=1                                                          ADPT 201
      CALL MESH2D      (MPOIN,MELEM,MNPEL,MBPLN,MPNOD,MDIME,MDENS,      ADPT 202
     -                  IADAP,IFRNT,NPOIN,NELEM,NNPEL,NMATR,NDENS,      ADPT 203
     -                  NBPLN,NBPOI,NBREF,NBELN,NBELF,NBELE,NCONC,      ADPT 204
     -                  MTYPE,NBELT,NBRET,NEWEL,NEWNO,COORD,XDENS,      ADPT 205
     -                  YDENS,EHNEW)                                    ADPT 206
      IF (ILINR.EQ.1)                                                  ADPT 207
     -CALL TRANSF       (MPOIN,MELEM,MNPEL,MDIME,MELE3,ICONV,ILINR,     ADPT 208
     -                   NPOIN,NNPEL,NPOIO,NELEO,NCONO,LNOD3,           ADPT 209
     -                   COORO,COORD,TEMPR,DAREP,TEMPA)                 ADPT 210
      GOTO 000                                                         ADPT 211
  999 CONTINUE                                                         ADPT 212
      STOP                                                             ADPT 213
C                                                                      ADPT 214
C *** PERFORM  TRANSIENT  ANALYSIS                                     ADPT 215
C                                                                      ADPT 216
  100 CONTINUE                                                         ADPT 217
  101 TIME=TIME+DTIME                                                  ADPT 218
      NSTEP=NSTEP+1                                                    ADPT 219
C                                                                      ADPT 220
C *** PLOTTING FILE                                                    ADPT 221
C                                                                      ADPT 222
      CALL TRANSI      (MELEM,MPOIN,MGAUS,MNPEL,MCDPT,MMATR,MTYPE,      ADPT 223
     -                  MDIME,MBOUN,MPROF,MFRON,IAXSY,ILINR,ITRAN,      ADPT 224
     -                  ICONV,IPETR,NITER,NFIXB,TOLER,DTMAX,FACTR,      ADPT 225
     -                  NELEM,NPOIN,NGASB,NGAUS,NNPEL,NITUP,NITDN,      ADPT 226
     -                  NNPFC,NPROF,NCDPT,NDFEL,NFACB,NBELN,NCONC,      ADPT 227
     -                  LNODS,NDIAG,IFFIX,TFIXD,TVALU,CDVLU,CPVLU,      ADPT 228
     -                  CONDY,CAPCY,TEMPT,SHAPF,DERV1,DERV2,ARWET,      ADPT 229
     -                  ASTIF,AMASS,POSGB,WEIGB,GSTIF,XCORD,YCORD,      ADPT 230
     -                  EFORC,FORCE,FLUXE,COEFF,RADIA,AMBIT,TLAST,      ADPT 231
     -                  TEMPR,COORD,UVELO,VVELO,ULOCA,VLOCA,PETRV,      ADPT 232
     -                  NDEST,LOCEL,LHEDV,NADFN,NODFN,PNORM,GFLUM,      ADPT 233
     -                  NSTEP,TTIME,ALPHA,DTIME,TEMP1,RVECT)            ADPT 234
C                                                                      ADPT 235
C*** CONDUCT ERROR ESTIMATION IF REQUIRED                              ADPT 236
C                                                                      ADPT 237
      IF(IERST.NE.0)                                                   ADPT 238
     -CALL ERREST       (MELEM,MGAUS,MNPEL,MCDPT,MMATR,MPOIN,NGAUS,     ADPT 239
     -                   NNPEL,ILINR,ICOAR,NCDPT,TVALU,CDVLU,SHAPF,     ADPT 240
     -                   DERV1,DERV2,NPOIN,MPROF,ARWET,MDIME,COORD,     ADPT 241
     -                   MTYPE,CONDY,TEMPR,NCONC,NELEM,GSTIF,IFFIX,     ADPT 242
     -                   FORCE,TENER,TGRDX,TGRDY,ERHS1,ERHS2,ERNOR,     ADPT 243
     -                   ERELN,SGRDX,SGRDY,SMOTH,MDIAG,NCOLM,PERMI,     ADPT 244
     -                   EHOLD,EHNEW,ICONT,ERDOM,ERSRT,ERMAX,KELRM,     ADPT 245
     -                   PCENT)                                         ADPT 246
C                                                                      ADPT 247
C *** OUTPUT RESULTS                                                   ADPT 248
C                                                                      ADPT 249
      WRITE (7,*) ' TIME = ',TIME, ' STEP = ',NSTEP                    ADPT 250
      CALL  OUTPUT      (MELEM,MPOIN,MMATR,MNPEL,NPOIN,NELEM,NNPEL,     ADPT 251
     .                   NMATR,IERST,NCONC,MTYPE,TEMPR,COORD,MDIME,     ADPT 252
     .                   ERELN,ERDOM,ERSRT,EHNEW,ERMAX,KELRM,UVELO,     ADPT 253
     .                   VVELO,TIME)                                    ADPT 254
C                                                                      ADPT 255
C *** IF REQUIRED ERROR NOT ATTAINED REMESH AND REANALYSE              ADPT 256
C                                                                      ADPT 257
      IF (ERSRT.LE.RCENT) GOTO 9999                                    ADPT 258
      NDENS=NPOIN                                                      ADPT 259
      DO I=1,NPOIN                                                     ADPT 260
```

```
         XDENS(I)=COORD(I,1)                                            ADPT 261
         YDENS(I)=COORD(I,2)                                            ADPT 262
      END DO                                                           ADPT 263
      DO I=1,NPOIN                                                     ADPT 264
         EHNEW(I)=EHNEW(I)*AMULT                                        ADPT 265
         IF (EHNEW(I).GT.ELMAX) EHNEW(I)=ELMAX                          ADPT 266
         IF (EHNEW(I).LT.ELMIN) EHNEW(I)=ELMIN                          ADPT 267
      END DO                                                           ADPT 268
C                                                                       ADPT 269
C *** TRANSFER OLD MESH INFO TO SEPERATE ARRAYS                         ADPT 270
C                                                                       ADPT 271
      NELEO=NELEM                                                       ADPT 272
      NPOIO=NPOIN                                                       ADPT 273
      DO I=1,NELEM                                                     ADPT 274
         DO J=1,NNPEL                                                  ADPT 275
            NCONO(I,J)=NCONC(I,J)                                       ADPT 276
         END DO                                                        ADPT 277
      END DO                                                           ADPT 278
      DO I=1,NPOIN                                                     ADPT 279
         COORO(I,1)=COORD(I,1)                                         ADPT 280
         COORO(I,2)=COORD(I,2)                                         ADPT 281
      END DO                                                           ADPT 282
      REWIND 11                                                        ADPT 283
      IADAP=1                                                          ADPT 284
      CALL MESH2D        (MPOIN,MELEM,MNPEL,MBPLN,MPNOD,MDIME,MDENS,   ADPT 285
     -                    IADAP,IFRNT,NPOIN,NELEM,NNPEL,NMATR,NDENS,   ADPT 286
     -                    NBPLN,NBPOI,NBREF,NBELN,NBELF,NBELE,NCONC,   ADPT 287
     -                    MTYPE,NBELT,NBRET,NEWEL,NEWNO,COORD,XDENS,   ADPT 288
     -                    YDENS,EHNEW)                                  ADPT 289
      CALL TRANSF        (MPOIN,MELEM,MNPEL,MDIME,MELE3,ICONV,ILINR,   ADPT 290
     -                    NPOIN,NNPEL,NPOIO,NELEO,NCONO,LNOD3,         ADPT 291
     -                    COORO,COORD,TEMPR,DAREP,TEMPA)               ADPT 292
      IF (TIME.GT.TTIME) STOP                                          ADPT 293
      GOTO 888                                                         ADPT 294
 9999 CONTINUE                                                         ADPT 295
      IF (TIME.LT.TTIME) GOTO 101                                      ADPT 296
      STOP                                                             ADPT 297
      END                                                              ADPT 298
```

B.4   Input Instructions

The input data required by this program can be divided into two distinct groups. The first group consists of data required for generating the mesh. This data is read from files **GEOMET.DAT** and **DENSIT.DAT** in various subroutines of the subprogram **MESH2D**. The control data and the material property and boundary condition data is read from file **AD-PINP.DAT** in subroutines **CONTRL** and **INDATA**. The data required for the first two files is explained first:

B.4.1   Geometry Data

In this section the data for the file **GEOMET.DAT** will be described. This data will be used for generating the finite element mesh of 3,4 or 6-noded elements. The variables to be input are written in bold face in bold brackets. All the variable names are as defined in the tables earlier.

1 Control data.

   1.1 **(nbnds,natrib,nnpe,toler)**

2 Internal regions data (*nbnds* times).

   2.1 **(noseg)**

   2.2 Segment details (for $i = 1$ to *noseg*) to be input in a clockwise order. Straight line (*ityp* $= 1$) segments are defined by two points. Any reference to the first point in a variable may be inferred from a '1' included in that variable. For referencing subsequent points the same procedure is followed. For arc segments (*ityp* $= 2$), a radius is required in addition to the two points. The radii of the circular arc segments are negative for arcs in a clockwise direction and positive for arcs in a counterclockwise direction. Direction is determined by moving from point 1 to point 2. The third type, a user defined segment (*ityp* $= 3$), can theoretically consist of any number of points.

     2.2.1 **(ityp)**

     2.2.2 For $i = 1$

       2.2.2.1 **(x1,y1,x2,y2)** (if *ityp* $= 1$)

       2.2.2.2 **(x1,y1,x2,y2,rad)** (if *ityp* $= 2$)

       2.2.2.3 **(numpt,xx(1),yy(1),....,xx(numpt),yy(numpt))** (if *ityp* $= 3$)

       2.2.2.4 **(iseg)**(if *ityp* $= 4$)

     2.2.3 For $i > 1$ and $i < noseg$

       2.2.3.1 **(x2,y2)** (if *ityp* $= 1$)

       2.2.3.2 **(x2,y2,rad)** (if *ityp* $= 2$)

       2.2.3.3 **(numpt,xx(2),yy(2),....,xx(numpt),yy(numpt))** (if *ityp* $= 3$)

       2.2.3.4 **(iseg)**(if *ityp* $= 4$)

     2.2.4 For $i = noseg$

       2.2.4.1 no value for this line (if *ityp* $= 1$)

       2.2.4.2 **(rad)** (if *ityp* $= 2$)

       2.2.4.3 **(numpt,xx(2),yy(2),....,xx(numpt-1),yy(numpt-1))** (if *ityp* $= 3$)

       2.2.4.4 **(iseg)**(if *ityp* $= 4$)

3 Attribute info for subdomains (*natrib* times).

   3.1 **(iatt(1),....,iatt(natrib))**

4 Subdomains data (*natrib* times). This is exactly the same as item 2 above except that this is input in a counter-clockwise order.

B.4.2 Mesh Density Data

In this section the data for the file **DENSIT.DAT** will be described. This is used to control the local element size (mesh density) for the first mesh only. This data simply consists of a list of points with a mesh density value attached to each point. Any number of such points may be defined by the user to control the element sizes in various regions of the mesh. The data is written as follows:

ndpt
xd(1),yd(1),vald(1)
.
.
.
xd(ndpt),yd(ndpt),vald(ndpt)

B.4.3 Example Data File

An arbitrary 2-D geometry shown in Figure B.2 will be used here to illustrate the preparation of data for the mesh generator as explained in the previous section. Only the numbers printed in typescript are the ones to be included in the datafiles.

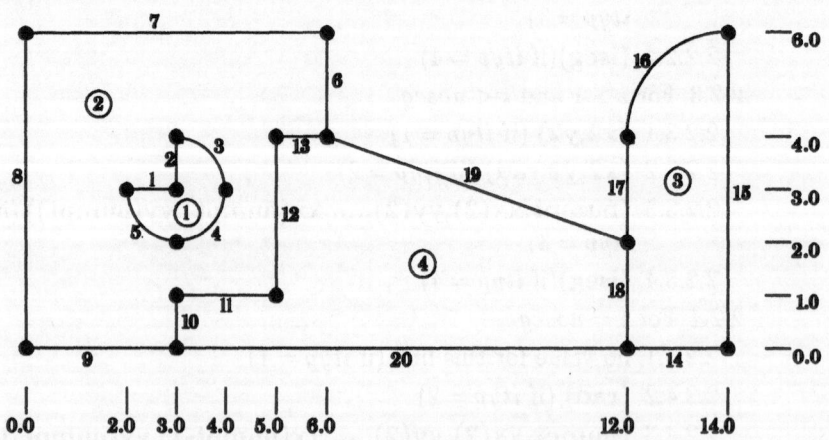

Figure B.2: Geometry for the example data file

Geometry Data

1 *Control data.*

```
1    3    3    0.00001
```

2 *Internal boundaries data* (*nbnds* times). In Figure B.2 the encircled numbers mark the subdomains. The encircled 1 marks the hole or the internal region, which can be seen to consist of 5 segments. The data required for these segments is written as below (according to the instructions earlier).

```
5
1
2.0 3.0 3.0 3.0
1
3.0 4.0
2
4.0 3.0 -1.0
2
3.0 2.0 -1.0
2
-1.
```

3 *Attribute info for subdomains.*

```
1,1,2
```

These numbers may be interpreted to mean that subdomains 2 and 3 consist of material 1 and subdomain 4 consists of material 2. All the elements generated in subdomains 2,3 and 4 will have the appropriate attribute numbers associated with them. Subdomain 1 is an internal region or a 'hole', therefore no elements will be generated there.

4 *External boundaries data* (*natrib* times). The input here is in a counter-clockwise order.

For subdomain 2,

```
8
1
6.  4.  6.  6.
1
```

```
0.  6.
1
0.  0.
1
3.  0.
1
3.  1.
1
5.  1.
1
5.  4.
1
```

For subdomain 3,

```
5
1
12.  0.  14.  0.
1
14.  6.
2
12.  4.  2.
1
12.  2.
1
```

For subdomain 4,

```
7
4
18
1
6.  4.
4
13
4
12
4
11
4
10
1
```

Mesh Density Data

```
5
0.0 0.0 2.0
0.0 6.0 2.0
3.0 3.0 0.4
7.0 2.0 2.0
14.0 4.0 1.0
```

B.4.4 Output Data

The mesh that results from the example data in the previous section is shown in Figure B.3. If the value of *nnpe* is changed to 4 in the data of the previous section the result is a mesh of 4-noded quadrilateral elements as shown in Figure B.4

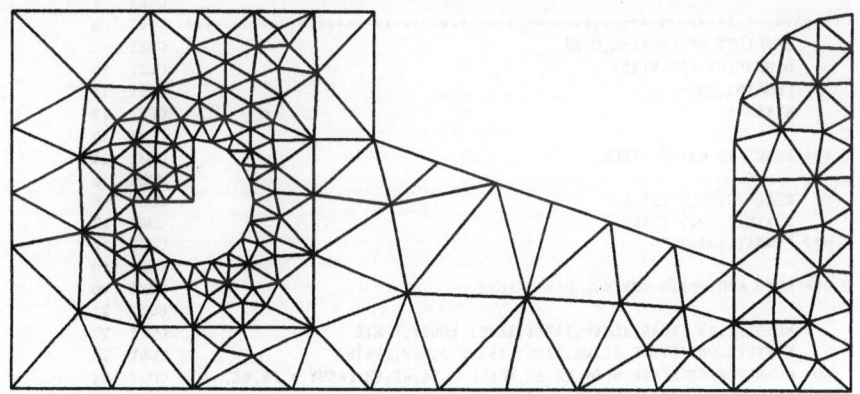

Figure B.3: Mesh of triangles for the example data file

B.4.5 Problem Data

In this section the data for the file **ADPINP.DAT** will be described. This data is read from the subroutines **CONTRL** and **INDATA** which are listed below.

```
        SUBROUTINE CONTRL (ITRAN,ILINR,IAXSY,IERST,ICONV,IPETR,NITER,    CONT  1
       -            NITUP,NITDN,TTIME,STIME,DTIME,DTMAX,ALPHA,           CONT  2
       -            RELAX,TOLER,PCENT,ELMIN,ELMAX,FACTR)                 CONT  3
C****************************************************************************  CONT  4
C                                                                        CONT  5
C *** THIS SUBROUTINE READS ALL PROBLEM DATA                             CONT  6
```

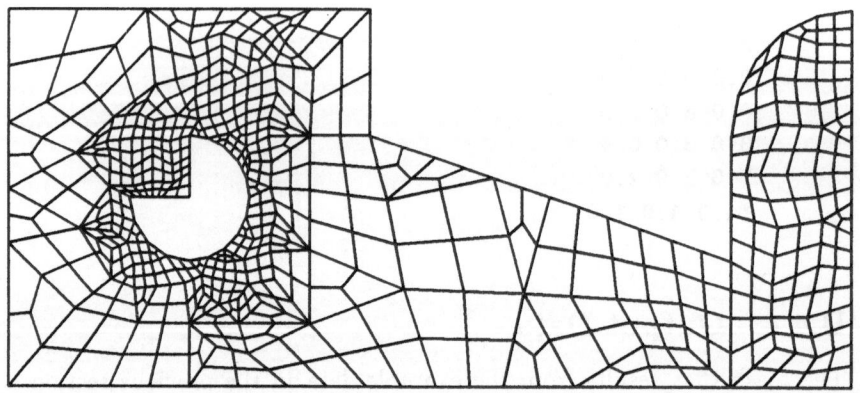

Figure B.4: Mesh of quadrilaterals for the example data file

```
C                                                                      CONT   7
C*********************************************************************** CONT   8
      IMPLICIT REAL*8(A-H,O-Z)                                         CONT   9
      DIMENSION TITLE(12)                                             CONT  10
      PCENT=1.0D0                                                     CONT  11
      NITER=1                                                          CONT  12
C                                                                      CONT  13
C *** READ AND WRITE TITLE                                            CONT  14
C                                                                      CONT  15
      READ(12,920) TITLE                                              CONT  16
      WRITE(7,920) TITLE                                              CONT  17
  920 FORMAT(12A6)                                                    CONT  18
C                                                                      CONT  19
C *** READ AND WRITE CONTROL PARAMETERS                               CONT  20
C                                                                      CONT  21
      READ(12,*) ITRAN,ILINR,IAXSY,IERST,ICONV,IPETR                  CONT  22
      WRITE(7,901)ITRAN,ILINR,IAXSY,IERST,ICONV,IPETR                 CONT  23
  901 FORMAT(//8H ITRAN =,I4,4X,8H ILINR =,I4,4X,8H IAXSY =,I4,4X,    CONT  24
     .8H IERST =,I4//8H ICONV =,I4,4X,8H IPETR =,I4)                  CONT  25
      IF (ITRAN.EQ.1) READ(12,*) TTIME,STIME,DTIME,ALPHA             CONT  26
      IF (ITRAN.EQ.1) WRITE(7,902)TTIME,STIME,DTIME,ALPHA            CONT  27
  902 FORMAT(//8H TTIME =,F5.3,3X,8H STIME =,F5.3,3X,                 CONT  28
     .8H DTIME =,F5.3,3X,8H ALPHA =,F5.3)                             CONT  29
      IF (ILINR.EQ.1) READ(12,*) NITER,TOLER,RELAX                   CONT  30
      IF (ILINR.EQ.1) WRITE(7,903) NITER,TOLER,RELAX                 CONT  31
  903 FORMAT(//8H NITER =,I4,4X,8H TOLER =,F5.3,3X,                   CONT  32
     .8H RELAX =,F5.3)                                                CONT  33
      IF (ILINR.EQ.1.AND.ITRAN.EQ.1) READ(12,*) NITUP,NITDN,DTMAX,FACTR CONT  34
      IF (ILINR.EQ.1.AND.ITRAN.EQ.1)WRITE(7,905) NITUP,NITDN,DTMAX,FACTRCONT 35
  905 FORMAT(//8H NITUP =,I4,4X,8H NITDN =,I4,4X,8H DTMAX =,F5.3,3X,   CONT  36
     .8H FACTR =,F5.3)                                                CONT  37
      IF (IERST.EQ.1) READ(12,*) PCENT,ELMIN,ELMAX                   CONT  38
      IF (IERST.EQ.1) WRITE(7,904) PCENT,ELMIN,ELMAX                 CONT  39
  904 FORMAT(//8H PCENT =,F5.3,3X,8H ELMIN =,F5.3,3X,3X,8H ELMAX =,F5.3)CONT 40
      RETURN                                                          CONT  41
      END                                                             CONT  42
```

All the problem data regarding the material and boundary conditions for the file **ADPINP.DAT** must be input as follows.

1 Control data (this item is read in subroutine CONTRL).

    1.1 **(title)**

    1.2 **(itran,ilinr,iaxsy,ierst,iconv,ipetr)**

    1.3 **(ttime,stime,dtime,alpha)** (Only if *itran*=1)

    1.4 **(niter,toler,relax)** (Only if *ilinr*=1)

    1.5 **(nitup,nitdn,dtmax,factr)** (Only if *ilinr*=1 and *itran*=1)

    1.6 **(pcent,elmin,elmax)** (Only if *ierst*=1)

2 Material properties (this and all subsequent items are read in subroutine INDATA)

    2.1 **(condy,capcy)** *nmatr* times

    2.2 **(ncdpt)** (Only if *ilinr*=1)
    As **HEAT2D** if *iphas*=0 (next line), and *itran*=1 and *ilinr*=1 then this program will also expect an enthalpy curve to be input.

    2.3 **(cdvlu,cpvlu,tvalu)** *ncdpt* times *nmatr* times

3 Dirichlet type boundary conditions

    3.1 **(nobcd)**

    3.2 **(tembc)** *nobcd* times

    3.3 **(num)** (Only if *nobcd*≠0)

    3.4 **(numb,numc)** *num* times
    Here if *numc* is positive then, the boundary condition is applied to the all the nodes of segment *numb*. If however, *numc* is negative, the boundary condition is applied only to the first node of segment *numb*.

4 Neumann type boundary conditions

    4.1 **(nobcn)**

    4.2 **(fluxp,coefp,radip,ambip)** *neumn* times

    4.3 **(num)** (Only if *nobcn*≠0)

    4.4 **(numb,numc)** *num* times Here if *numc* is positive then, the boundary condition is applied to the all the elements of segment *numb*. If however, *numc* is negative, the boundary condition is applied only to the first element of segment *numb*.

5 Initial conditions

    5.1 **(temin)** *nmatr* times (Only if *itran*=1 or *ilinr*=1)   .

6 Velocity field

    6.1 **(uvelm,vvelm)** *nmatr* times (Only if *iconv*=1)

B.5   Error Estimate Calculations

Error estimates are calculated in the following routine.

```
      SUBROUTINE ERREST (MELEM,MGAUS,MMPEL,MCDPT,MMATR,MPOIN,NGAUS,     ERRE   1
     -            NNPEL,ILINR,ICOAR,NCDPT,TVALU,CDVLU,SHAPF,            ERRE   2
     -            DERV1,DERV2,NPOIN,MPROF,ARWET,NDIME,COORD,            ERRE   3
     -            MTYPE,CONDY,TEMPR,NCONC,NELEM,GMASS,IFFIX,            ERRE   4
     -            FORCE,TENER,TGRDX,TGRDY,ERHS1,ERHS2,ERNOR,            ERRE   5
     -            ERELM,SGRDX,SGRDY,SMOTH,NDIAG,NCOLM,PERMI,            ERRE   6
     -            EHOLD,EHNEW,ICONT,ERDOM,ERSRT,ERMAX,KELRM,            ERRE   7
     -            PCENT)                                                ERRE   8
C**********************************************************************  ERRE   9
C                                                                       ERRE  10
C**** SUBROUTINE TO PERFORM ERROR ESTIMATION CALCULATIONS               ERRE  11
C                                                                       ERRE  12
C**********************************************************************  ERRE  13
C     INSERT DOUBLE                                                     ERRE  14
C                                                                       ERRE  15
      IMPLICIT REAL*8(A-H,O-Z)                                          ERRE  16
      DIMENSION SHAPF(MNPEL,MGAUS,MELEM),DERV1(MNPEL,MGAUS,MELEM),      ERRE  17
     -          DERV2(MNPEL,MGAUS,MELEM),ARWET(MGAUS,MELEM),            ERRE  18
     -          CDVLU(MCDPT,MMATR),NDIAG(MPOIN),NCONC(MELEM,MNPEL),     ERRE  19
     -          TVALU(MCDPT,MMATR),TEMPR(MPOIN),COORD(MPOIN,MDIME),     ERRE  20
     -          MTYPE(MELEM),CONDY(MMATR),GMASS(MPROF),IFFIX(MPOIN),    ERRE  21
     -          SHAPP(9),CARTP(2,9)                                     ERRE  22
      DIMENSION TENER(MELEM),TGRDX(MELEM,MGAUS),TGRDY(MELEM,MGAUS),     ERRE  23
     .          ERHS1(MELEM,MNPEL),ERHS2(MELEM,MNPEL),NCOLM(MPOIN),     ERRE  24
     .          ERNOR(MELEM),SGRDX(MELEM,MGAUS),SGRDY(MELEM,MGAUS),     ERRE  25
     .          ERELM(MELEM),FORCE(MPOIN),SMOTH(MPOIN),                 ERRE  26
     .          PERMI(MELEM),EMASS(9,9),XE(3),YE(3),FX(9),FY(9),        ERRE  27
     .          EHOLD(MELEM),EHNEW(MPOIN),ICONT(MPOIN)                  ERRE  28
      DIMENSION IJ4(3),IJ9(3)                                          ERRE  29
      DATA IJ4/1,3,4/,IJ9/1,5,7/                                        ERRE  30
      SIZEO=0.0                                                         ERRE  31
C                                                                       ERRE  32
C *** IF ELEMENTS ALLOWED TO BE COARSENED, ICOAR=1                      ERRE  33
C                                                                       ERRE  34
C     ICOAR=1                                                           ERRE  35
C                                                                       ERRE  36
C *** CALCULATE ELEMENT SIZES OF THE PRESENT MESH                       ERRE  37
C                                                                       ERRE  38
      ESIZE=0.0                                                         ERRE  39
      IF(NNPEL.EQ.3.OR.NNPEL.EQ.6) THEN                                 ERRE  40
      DO 1010 IEL=1,NELEM                                               ERRE  41
      DO 1015 JPE=1,3                                                   ERRE  42
      JJ=JPE                                                            ERRE  43
      IF(NNPEL.EQ.6) JJ=JPE*2-1                                         ERRE  44
      NP=NCONC(IEL,JJ)                                                  ERRE  45
      XE(JPE)=COORD(NP,1)                                               ERRE  46
 1015 YE(JPE)=COORD(NP,2)                                               ERRE  47
      A1=SQRT((XE(2)-XE(1))**2+(YE(2)-YE(1))**2)                        ERRE  48
      A2=SQRT((XE(3)-XE(2))**2+(YE(3)-YE(2))**2)                        ERRE  49
```

```
      A3=SQRT((XE(1)-XE(3))**2+(YE(1)-YE(3))**2)         ERRE 50
      A0=0.5*(A1+A2+A3)                                  ERRE 51
      AA=A0*(A0-A1)*(A0-A2)*(A0-A3)                      ERRE 52
      AS=SQRT(AA)*2.0                                    ERRE 53
      ASIZ=SQRT(AS)                                      ERRE 54
      ESIZE=ESIZE+AS*0.5                                 ERRE 55
 1010 EHOLD(IEL)=ASIZ                                    ERRE 56
      END IF                                             ERRE 57
C                                                        ERRE 58
      IF(NNPEL.EQ.4.OR.NNPEL.EQ.9) THEN                  ERRE 59
      DO 1012 IEL=1,NELEM                                ERRE 60
      DO 1025 JPE=1,3                                    ERRE 61
      JJ=JPE                                             ERRE 62
      IF(NNPEL.EQ.9) JJ=JPE*2-1                          ERRE 63
      NP=NCONC(IEL,JJ)                                   ERRE 64
      XE(JPE)=COORD(NP,1)                                ERRE 65
 1025 YE(JPE)=COORD(NP,2)                                ERRE 66
      A1=SQRT((XE(2)-XE(1))**2+(YE(2)-YE(1))**2)         ERRE 67
      A2=SQRT((XE(3)-XE(2))**2+(YE(3)-YE(2))**2)         ERRE 68
      A3=SQRT((XE(1)-XE(3))**2+(YE(1)-YE(3))**2)         ERRE 69
      A0=0.5*(A1+A2+A3)                                  ERRE 70
      AA=A0*(A0-A1)*(A0-A2)*(A0-A3)                      ERRE 71
      AS1=SQRT(AA)                                       ERRE 72
C                                                        ERRE 73
      DO 1035 JPE=1,3                                    ERRE 74
      JJ=IJ4(JPE)                                        ERRE 75
      IF(NNPEL.EQ.9) JJ=IJ9(JPE)                         ERRE 76
      NP=NCONC(IEL,JJ)                                   ERRE 77
      XE(JPE)=COORD(NP,1)                                ERRE 78
 1035 YE(JPE)=COORD(NP,2)                                ERRE 79
      A1=SQRT((XE(2)-XE(1))**2+(YE(2)-YE(1))**2)         ERRE 80
      A2=SQRT((XE(3)-XE(2))**2+(YE(3)-YE(2))**2)         ERRE 81
      A3=SQRT((XE(1)-XE(3))**2+(YE(1)-YE(3))**2)         ERRE 82
      A0=0.5*(A1+A2+A3)                                  ERRE 83
      AA=A0*(A0-A1)*(A0-A2)*(A0-A3)                      ERRE 84
      AS2=SQRT(AA)                                       ERRE 85
      AS=AS1+AS2                                         ERRE 86
      ASIZ=SQRT(AS)                                      ERRE 87
      ESIZE=ESIZE+AS                                     ERRE 88
 1012 EHOLD(IEL)=ASIZ                                    ERRE 89
      END IF                                             ERRE 90
C                                                        ERRE 91
      ESIZE=SQRT(ESIZE)                                  ERRE 92
C                                                        ERRE 93
C *** SET UP THE VECTOR 'NDIAG' STORING THE DIAGONAL DOF NUMBER FOR  ERRE 94
C *** EACH COLUMN OF THE GLOBAL SMOOTHING MTX            ERRE 95
C                                                        ERRE 96
      CALL DIAGNL      (NELEM,NNPEL,NELEN,NPOIN,NNPEL,MPOIN,NCOLN,  ERRE 97
     -                  NCONC,NDIAG,NPROF)               ERRE 98
      DO 121 I=1,NPROF                                   ERRE 99
      GMASS(I)=0.0                                       ERRE 100
  121 CONTINUE                                           ERRE 101
      DO 5 IEL=1,NELEM                                   ERRE 102
      TENER(IEL)=0.0                                     ERRE 103
      ERROR(IEL)=0.0                                     ERRE 104
      DO 5 I3=1,NGAUS                                    ERRE 105
      TGRDX(IEL,I3)=0.0                                  ERRE 106
    5 TGRDY(IEL,I3)=0.0                                  ERRE 107
C                                                        ERRE 108
C *** START SMOOTHING PROCEDURE                          ERRE 109
C                                                        ERRE 110
      FINT=0.0                                           ERRE 111
      ELN=NELEM                                          ERRE 112
      DO 10 IEL=1,NELEM                                  ERRE 113
      DO 15 I1=1,NGAUS                                   ERRE 114
      FX(I1)=0.0                                         ERRE 115
   15 FY(I1)=0.0                                         ERRE 116
```

```
        DO 130 IPE=1,NNPEL                                           ERRE 117
        DO 130 JPE=1,NNPEL                                           ERRE 118
  130   EMASS(IPE,JPE)=0.0                                           ERRE 119
        IMATR=NTYPE(IEL)                                             ERRE 120
        CONDT=CONDY(IMATR)                                           ERRE 121
C                                                                    ERRE 122
C *** EXTRACT ELEMENT SHAPE FUNCTIONS                                ERRE 123
C                                                                    ERRE 124
        DO 20 IGAUS=1,NGAUS                                          ERRE 125
        GWIG=ARWET(IGAUS,IEL)                                        ERRE 126
          DO 32 INODP=1,NNPEL                                        ERRE 127
            SHAPP(INODP)=SHAPF(INODP,IGAUS,IEL)                      ERRE 128
            CARTP(1,INODP)=DERV1(INODP,IGAUS,IEL)                    ERRE 129
            CARTP(2,INODP)=DERV2(INODP,IGAUS,IEL)                    ERRE 130
   32     CONTINUE                                                   ERRE 131
C                                                                    ERRE 132
C *** EVALUATE PROPERTIES AT INTEGRATION POINTS                      ERRE 133
C                                                                    ERRE 134
        COND=CONDT                                                   ERRE 135
        IF (ILINR.EQ.1) THEN                                         ERRE 136
          TEMP =0.0                                                  ERRE 137
          DO 52 IPE=1,NNPEL                                          ERRE 138
            NP=NCONC(IEL,IPE)                                        ERRE 139
            TEMP=TEMP+TEMPR(NP)*SHAPP(IPE)                           ERRE 140
   52     CONTINUE                                                   ERRE 141
          CALL NONLIN (NCDPT,NMATR,IMATR,NCDPT,TEMP,TVALU,CDVLU,VALUE)ERRE 142
          COND=COND*VALUE                                            ERRE 143
        END IF                                                       ERRE 144
        CCCX=COND                                                    ERRE 145
        CCCY=COND                                                    ERRE 146
C                                                                    ERRE 147
C *** CALCULATE TEMPERATURE GRADIENTS FROM FEM RESULTS               ERRE 148
C                                                                    ERRE 149
        DO 30 IPE=1,NNPEL                                            ERRE 150
        NP=NCONC(IEL,IPE)                                            ERRE 151
        TGRDX(IEL,IGAUS)=TGRDX(IEL,IGAUS)+CARTP(1,IPE)*TEMPR(NP)     ERRE 152
   30   TGRDY(IEL,IGAUS)=TGRDY(IEL,IGAUS)+CARTP(2,IPE)*TEMPR(NP)     ERRE 153
        CCCXW=CCCX*GWIG                                              ERRE 154
        CCCYW=CCCY*GWIG                                              ERRE 155
C                                                                    ERRE 156
C *** OBTAIN SQUARE OF TOTAL HEAT DISSIPATION AND SMOOTHING FORCES    ERRE 157
C                                                                    ERRE 158
        TENER(IEL)=TENER(IEL)+CCCXW*TGRDX(IEL,IGAUS)*TGRDX(IEL,IGAUS) ERRE 159
                +CCCYW*TGRDY(IEL,IGAUS)*TGRDY(IEL,IGAUS)             ERRE 160
        DO 25 IPE=1,NNPEL                                           ERRE 161
        GDS=GWIG*SHAPP(IPE)                                          ERRE 162
        FX(IPE)=FX(IPE)+TGRDX(IEL,IGAUS)*GDS                         ERRE 163
   25   FY(IPE)=FY(IPE)+TGRDY(IEL,IGAUS)*GDS                         ERRE 164
C                                                                    ERRE 165
C *** CALCULATE MASS MATRIX                                          ERRE 166
C                                                                    ERRE 167
        DO 55 IPE=1,NNPEL                                            ERRE 168
        DO 55 JPE=1,NNPEL                                            ERRE 169
        EMASS(IPE,JPE)=EMASS(IPE,JPE)+GWIG*SHAPP(IPE)*SHAPP(JPE)     ERRE 170
   55 CONTINUE                                                       ERRE 171
   20 CONTINUE                                                       ERRE 172
C                                                                    ERRE 173
C *** ASSEMBLE INTO GLOBAL MASS MATRIX IN VECTOR FORM                ERRE 174
C                                                                    ERRE 175
          DO 128 J=1,NNPEL                                           ERRE 176
            DO 129 K=J,NNPEL                                         ERRE 177
              IROW=NCONC(IEL,J)                                      ERRE 178
              ICOL=NCONC(IEL,K)                                      ERRE 179
              IF (IROW.LE.ICOL) GOTO 104                             ERRE 180
              ITEM=IROW                                              ERRE 181
              IROW=ICOL                                              ERRE 182
              ICOL=ITEM                                              ERRE 183
```

```
104            IRC=NDIAG(ICOL)-ICOL+IROW          ERRE 184
               GMASS(IRC)=GMASS(IRC)+EMASS(J,K)   ERRE 185
103          CONTINUE                             ERRE 186
129        CONTINUE                               ERRE 187
128      CONTINUE                                 ERRE 188
      DO 26 IPE=1,NNPEL                           ERRE 189
      ERHS1(IEL,IPE)=FX(IPE)                      ERRE 190
   26 ERHS2(IEL,IPE)=FY(IPE)                      ERRE 191
      FINT=FINT+TENER(IEL)                        ERRE 192
   10 CONTINUE                                    ERRE 193
C                                                 ERRE 194
C *** HEAT DISSIPATIONAND IN EACH ELEMENT         ERRE 195
C                                                 ERRE 196
      FELEM=FINT/ELN                              ERRE 197
      FSRT=SQRT(FELEM)                            ERRE 198
C                                                 ERRE 199
C *** SOLVE DT/DX FOR SMOOTHED VALUES AT NODAL POINTS  ERRE 200
C                                                 ERRE 201
         DO 12 I=1,NPOIN                          ERRE 202
            FORCE(I)=0.0                          ERRE 203
   12    CONTINUE                                 ERRE 204
      DO 35 IEL=1,NELEM                           ERRE 205
      DO 35 IPE=1,NNPEL                           ERRE 206
      NOD=NCONC(IEL,IPE)                          ERRE 207
      FORCE(NOD)=FORCE(NOD)+ERHS1(IEL,IPE)        ERRE 208
   35 CONTINUE                                    ERRE 209
      DO 21 IPOIN=1,NPOIN                         ERRE 210
   21 SMOTH(IPOIN)=0.0                            ERRE 211
      CALL PROFAC (MPROF,MPOIN,NPOIN,GMASS,NDIAG) ERRE 212
      CALL PROSOL (MPROF,MPOIN,NPOIN,GMASS,FORCE,SMOTH,NDIAG)  ERRE 213
      DO 60 IEL=1,NELEM                           ERRE 214
      DO 62 IGAUS=1,NGAUS                         ERRE 215
      AA=0.0                                      ERRE 216
      DO 65 IPE=1,NNPEL                           ERRE 217
         NP=NCONC(IEL,IPE)                        ERRE 218
         SHAPP(IPE)=SHAPF(IPE,IGAUS,IEL)          ERRE 219
   65 AA=AA+SHAPP(IPE)*SMOTH(NP)                  ERRE 220
   62 SGRDX(IEL,IGAUS)=AA                         ERRE 221
   60 CONTINUE                                    ERRE 222
C                                                 ERRE 223
C *** SOLVE DT/DY FOR SMOOTHED VALUES AT NODAL POINTS  ERRE 224
C                                                 ERRE 225
         DO 42 I=1,NPOIN                          ERRE 226
            FORCE(I)=0.0                          ERRE 227
   42    CONTINUE                                 ERRE 228
      DO 45 IEL=1,NELEM                           ERRE 229
      DO 45 IPE=1,NNPEL                           ERRE 230
      NOD=NCONC(IEL,IPE)                          ERRE 231
      FORCE(NOD)=FORCE(NOD)+ERHS2(IEL,IPE)        ERRE 232
   45 CONTINUE                                    ERRE 233
      DO 22 IPOIN=1,NPOIN                         ERRE 234
   22 SMOTH(IPOIN)=0.0                            ERRE 235
      CALL PROSOL (MPROF,MPOIN,NPOIN,GMASS,FORCE,SMOTH,NDIAG)  ERRE 236
      DO 70 IEL=1,NELEM                           ERRE 237
      DO 72 IGAUS=1,NGAUS                         ERRE 238
      BB=0.0                                      ERRE 239
      DO 75 IPE=1,NNPEL                           ERRE 240
         NP=NCONC(IEL,IPE)                        ERRE 241
         SHAPP(IPE)=SHAPF(IPE,IGAUS,IEL)          ERRE 242
   75 BB=BB+SHAPP(IPE)*SMOTH(NP)                  ERRE 243
   72 SGRDY(IEL,IGAUS)=BB                         ERRE 244
   70 CONTINUE                                    ERRE 245
C                                                 ERRE 246
C *** START ERROR ESTIMATION                      ERRE 247
C                                                 ERRE 248
      ERMAX=0.0                                   ERRE 249
      EINT=0.0                                    ERRE 250
```

```
        DO 90 IEL=1,NELEM                                         ERRE 251
            IMATR=NTYPE(IEL)                                      ERRE 252
            CONDT=CONDY(IMATR)                                    ERRE 253
        DO 95 IGS=1,NGAUS                                         ERRE 254
        GWIG=ARWET(IGS,IEL)                                       ERRE 255
C                                                                 ERRE 256
C *** EVALUATE PROPERTIES AT INTEGRATION POINTS                   ERRE 257
C                                                                 ERRE 258
            COND=CONDT                                            ERRE 259
            IF (ILINR.EQ.1) THEN                                  ERRE 260
            TEMP =0.0                                             ERRE 261
            DO 402 IPE=1,NNPEL                                    ERRE 262
                NP=NCONC(IEL,IPE)                                 ERRE 263
                TEMP=TEMP+TEMPR(NP)*SHAPP(IPE)                    ERRE 264
    402     CONTINUE                                              ERRE 265
            CALL NONLIN (NCDPT,NMATR,IMATR,NCDPT,TEMP,TVALU,CDVLU,VALUE)ERRE 266
            COND=COND*VALUE                                       ERRE 267
            END IF                                                ERRE 268
C                                                                 ERRE 269
C *** CALCULATE THE NORM ERROR IN EACH ELEMENT                    ERRE 270
C                                                                 ERRE 271
        CCCX=COND                                                 ERRE 272
        CCCY=COND                                                 ERRE 273
        WJ=GWIG                                                   ERRE 274
        DELX=SGRDX(IEL,IGS)-TGRDX(IEL,IGS)                        ERRE 275
        DELY=SGRDY(IEL,IGS)-TGRDY(IEL,IGS)                        ERRE 276
     95 ERNOR(IEL)=ERNOR(IEL)+CCCX*WJ*DELX*DELX+CCCY*WJ*DELY*DELY ERRE 277
        ERELM(IEL)=SQRT(ERNOR(IEL))/FSRT                          ERRE 278
        IF(ERELM(IEL).GT.ERMAX) THEN                              ERRE 279
        ERMAX=ERELM(IEL)                                          ERRE 280
        KELRM=IEL                                                 ERRE 281
        END IF                                                    ERRE 282
C                                                                 ERRE 283
C *** CALCULATE XI AND NEW ELEMENT SIZES                          ERRE 284
C                                                                 ERRE 285
        ARR=PCENT/ERELM(IEL)                                      ERRE 286
        IF(NNPEL.EQ.6.OR.NNPEL.EQ.9) ARR=SQRT(ARR)                ERRE 287
        IF(ICOAR.NE.0) GO TO 92                                   ERRE 288
        IF(ARR.GT.1.0) ARR=1.0                                    ERRE 289
     92 PERMI(IEL)=EHOLD(IEL)*ARR                                 ERRE 290
        EINT=EINT+ERNOR(IEL)                                      ERRE 291
     90 CONTINUE                                                  ERRE 292
C                                                                 ERRE 293
C *** FIND THE AVERAGE NORM ERROR IN WHOLE DOMAIN                 ERRE 294
C                                                                 ERRE 295
        ERDOM=EINT/FINT                                           ERRE 296
C                                                                 ERRE 297
C *** AND CORRESPONDING TEMPERATURE ERROR IN WHOLE DOMAIN         ERRE 298
C                                                                 ERRE 299
        ERSRT=SQRT(ERDOM)                                         ERRE 300
C                                                                 ERRE 301
C *** AVERAGE ELEMENT SIZES AROUND NODAL POINTS                   ERRE 302
C                                                                 ERRE 303
        DO 110 IN=1,NPOIN                                         ERRE 304
        ICONT(IN)=0                                               ERRE 305
    110 EHNEW(IN)=0.0                                             ERRE 306
        DO 115 IEL=1,NELEM                                        ERRE 307
        DO 115 IPE=1,NNPEL                                        ERRE 308
            NP=NCONC(IEL,IPE)                                     ERRE 309
        ICONT(NP)=ICONT(NP)+1                                     ERRE 310
    115 EHNEW(NP)=EHNEW(NP)+PERMI(IEL)                            ERRE 311
        DO 120 IN=1,NPOIN                                         ERRE 312
        CONI=ICONT(IN)                                            ERRE 313
    120 EHNEW(IN)=EHNEW(IN)/CONI                                  ERRE 314
C                                                                 ERRE 315
C *** ADJUST ELEMENT SIZES WITHIN MINIMUM AND MAXIMUM             ERRE 316
C                                                                 ERRE 317
```

```
      DO 210 INP=1,NPOIN                                 ERRE 318
      IF(EHNEW(INP).LT.SIZEO) EHNEW(INP)=SIZEO           ERRE 319
      IF(EHNEW(INP).GT.ESIZE) EHNEW(INP)=ESIZE           ERRE 320
  210 CONTINUE                                           ERRE 321
      RETURN                                             ERRE 322
      END                                                ERRE 323
```

B.6 Documented Examples

In this section three examples of adaptive heat transfer analysis are presented with the the details of input. The output files for each example may be found in the floppy disk.

B.6.1 2-D Heat Conduction with Convective Boundary Condition

The first example is the same as used in Chapter 6 to demonstrate the adaptive analysis technique. Figure B.5 shows the geometry and boundary condition details required to prepare the input data files for this example.

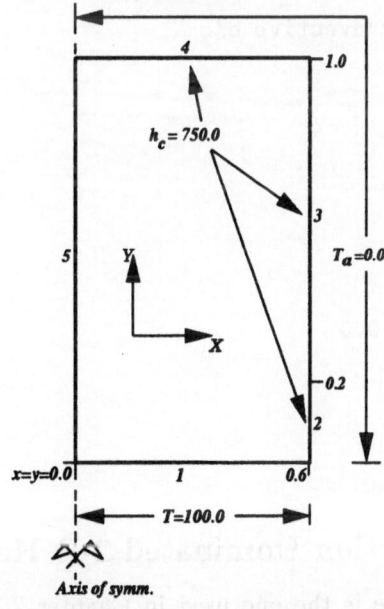

Figure B.5: Geometry and BC details for the 2-D heat conduction example

The first data file to be prepared is **GEOMET.DAT**. For this example it looks like this:

```
0 1 4 0.000001
```

```
1
5
1
0.0 0.0 0.6 0.0
1
0.6 0.2
1
0.6 1.0
1
0.0 1.0
1
```

The second data file to be prepared is **DENSIT.DAT**, which for this example is:

```
1
0.0 0.0 0.4
```

The third and final data file to be prepared is **ADPINP.DAT**, which for this example is:

```
Conduction with convective b/c
0  0  0  1  0  0
0.10,0.02,0.5
52.0      1.0
1
100.0
1
1 1
1
0.0  750.0  0.0  0.0
3
2 1
3 1
4 1
```

B.6.2 Convection Dominated 2-D Heat Transfer

The second example is the one used in Chapter 7 to demonstrate the Petrov-Galerkin method for steady-state convection dominated problems. Figure B.6 shows the geometry and boundary condition details required to prepare the input data files for this example.

 GEOMET.DAT for this example is:

```
0  1  3  0.000001
1
```

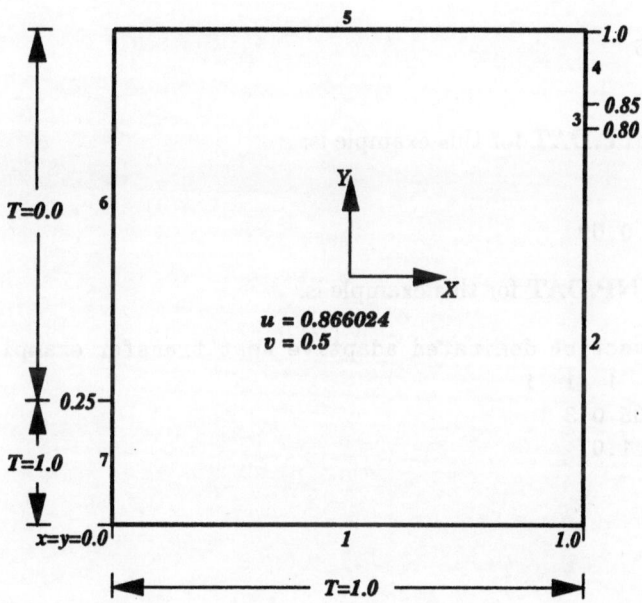

Figure B.6: Geometry and BC details of the convection dominated 2-D heat transfer example

```
7
1
0.0 0.0 1.0 0.0
1
1.0 0.8
1
1.0 0.85
1
1.0 1.0
1
0.0 1.0
1
0.0 0.25
1
```

DENSIT.DAT for this example is:

```
1
0.0 0.0 0.09
```

ADPINP.DAT for this example is:

```
2-D convection dominated adaptive heat transfer example
0   0   0   1   1   1
0.2 0.005 0.3
1.0d-6  1.0
2
1.0
0.0
3
1 1
6 2
7 1
0
0.8660254  0.5
```

B.6.3   1-D Solidification Example

The third example is the same as the first example of Appendix A. However, it has been solved adaptively here using three noded triangular elements with the enthalpy method to account for the latent heat effects instead of the source method used earlier.

GEOMET.DAT for this example is:

```
0 1 3 0.000001
1
```

```
6
1
0.0 0.0 1.0 0.0
1
4.0 0.0
1
4.0 0.5
1
1.0 0.5
1
0.0 0.5
1
```

DENSIT.DAT for this example is:

```
8
0.0 0.0 0.05
0.5 0.0 0.1
1.0 0.0 0.2
4.0 0.0 0.5
0.0 0.5 0.05
0.5 0.5 0.1
1.0 0.5 0.2
4.0 0.5 0.5
```

ADPINP.DAT for this example is:

```
1-D Phase change adaptive example data
1 1 0 1 0 0
4.0 0.0 0.000001 0.5
9 0.01 1.0
3 5  0.1 1.5
0.30  0.02 0.5
1.08 1.0
4
1.0   1.0  -50.0
1.0  49.85 -1.15
1.0 121.11 -0.15
1.0 126.26  5.0
1
-45.0
1
6 1
0
0.0
```

It may be noted that the values of *itran* and *ilinr* for this data file are 1. This means that the program expects an enthalpy curve to take account of the heat capacity. This allows us to model phase change problems using the enthalpy method. The values of *condy* and *capcy* here are the same as the example of Appendix A. The difference is that the latent heat value of 70.26 is incorporated in the enthalpy curve. The relationship between enthalpy and heat capacity is given in Equations (5.5) and (5.6), and is schematically shown in Figure 5.1. From this information we can easily construct an enthalpy curve for this problem. We begin with a temperature value of -50.0 and a corresponding arbitrarily chosen value of enthalpy (H) as 1 (as only the quantity $\frac{dH}{dT}$ is of interst and not the absolute value of H). Assuming a temperature range of -1.15 to -0.15 for solidification and going on to a suitable last value of 5.0 we determine all our temperature points. The corresponding enthalpy values are obtained by integrating the heat capacity between these temperature ranges. The second value is obtained by:

$$(-1.15 - (-50.0)) \times 1 + 1 = 49.85$$

For the third value we add the latent heat as:

$$(-0.15 - (-1.15)) \times 1 + 49.85 + 70.26 = 121.11$$

and similarly the fourth value is determined.

Subject Index